Also by Lori Andrews

Genetics: Ethics, Law and Policy (with Mark Rothstein and Maxwell Mehlman)

Immunity

The Silent Assassin

Sequence

Future Perfect: Confronting Decisions About Genetics

Body Bazaar: The Market for Human Tissue in the Biotechnology Age (with Dorothy Nelkin)

The Clone Age: Adventures in the New World of Reproductive Technology

Black Power, White Blood: The Life and Times of Johnny Spain

Assessing Genetic Risks: Implications for Health and Social Policy (co-edited with Jane E. Fullarton, Neil A. Holtzman, and Arno G. Motulsky)

Between Strangers: Surrogate Mothers, Expectant Fathers, and Brave New Babies

Medical Genetics: A Legal Frontier

New Conceptions: A Consumer's Guide to the Newest Infertility Treatments, Including In Vitro Fertilization, Artificial Insemination, and Surrogate Motherhood

I KNOW WHO YOU ARE AND I SAW WHAT YOU DID

Social Networks and the Death of Privacy

Lori Andrews

Free Press

New York   London   Toronto   Sydney   New Delhi

Free Press
A Division of Simon & Schuster, Inc.
1230 Avenue of the Americas
New York, NY 10020

First Free Press hardcover edition January 2012

FREE PRESS and colophon are trademarks of Simon & Schuster, Inc.

For information about special discounts for bulk purchases,
please contact Simon & Schuster Special Sales at
1-866-506-1949 or business@simonandschuster.com.

The Simon & Schuster Speakers Bureau can bring authors to your live event.
For more information or to book an event contact the Simon & Schuster Speakers Bureau
at 1-866-248-3049 or visit our website at www.simonspeakers.com.

Designed by Jason Snyder

Manufactured in the United States of America

10 9 8 7 6 5 4 3 2 1

Library of Congress Cataloging-in-Publication Data
Andrews, Lori B.
    I know who you are and I saw what you did : social networks and the death of privacy / by
Lori Andrews.
        p.   cm.
    Includes bibliographical references and index.
    1. Internet—Political aspects. 2. Internet—Law and legislation. 3. Online social networks—
Political aspects. 4. Privacy, Right of. 5. Civil rights. I. Title.
HM851.A66 2012
323.0285'4678—dc23
                                                                            2011034191

ISBN 978-1-4516-5051-8
ISBN 978-1-4516-5106-5 (ebook)

To the People of the Facebook/Twitter/
Google/YouTube/Myspace Nation—

May your data never be used against you

CONTENTS

I KNOW WHO YOU ARE
AND I SAW WHAT YOU DID

Facebook Nation

When David Cameron became Britain's prime minister, he made an appointment to talk to another head of state — Mark Zuckerberg. Yes, that Mark Zuckerberg: the billionaire wunderkind, the founder of Facebook. At the meeting at 10 Downing Street, Prime Minister Cameron and Facebook President Zuckerberg discussed ways in which social networks could take over certain governmental duties and inform public policymaking.[1]

A month later, Zuckerberg and Cameron had a follow-up conversation, later posted on YouTube. Cameron, dressed in suit and tie, chatted with Zuckerberg, who wore a blue cotton T-shirt.[2] "Basically, we've got a big problem here," Cameron pointed out to Zuckerberg, describing the U.K.'s financial woes.

Zuckerberg outlined how Facebook could be used as a platform to decrease spending and increase public participation in the political process: "I mean all these people have great ideas and a lot of energy that they want to bring and I think for a lot of people it's just about having an easy and a cheap way for them too to communicate their ideas."

"Brilliant," Cameron said.

Within a year, Zuckerberg had a seat at the table with government leaders. In May 2011, he attended the G8 Summit, the annual meeting of key heads of state (named after the eight advanced economies — France, the United States, the United Kingdom, Germany, Japan, Italy, Canada, and Russia).[3] The media reported that world leaders from German Chancellor Angela Merkel to French President Nicolas Sarkozy were more in awe of Zuckerberg than he was of them.[4] Zuckerberg summarized how Facebook had played a role in worldwide democratic movements and pressed his own policy agenda — urging European officials to back off of proposed regulation of the internet. "People tell me on the one hand 'It's great you played such a big role in the Arab spring, but it's also kind of scary because you enable all this sharing and collect information on people,'" Zuckerberg said.

Is it odd to think of Mark Zuckerberg as a head of state? Perhaps. But Facebook has the power and reach of a nation. With more than 750 million members, Facebook's population would make it the third largest nation in the world. It has citizens, an economy, its own currency, systems for resolving disputes, and relations with other nations and institutions. After watching the video chat between Cameron and Zuckerberg, I became intrigued by the concept of a social network as a nation. I began to wonder, what kind of government rules Facebook? What are its politics? And, if it is like a nation, should it have a Constitution?

People are drawn to Facebook, as early settlers are drawn to any new nation, by the search for freedom. Social networks expand people's opportunities. An ordinary individual can be a reporter, alerting the world to breaking news of a natural disaster or a political crisis. Or an investigator, helping cops solve a crime. Filmmakers and musicians at the start of their careers can find large followings through social networks.

The power of people is harnessed in new ways on social networks. Art itself is redefined as bands and novelists post early works and use crowdsourcing to change the music, lyrics, and story lines. Anybody can be a scientist, participating in a crowdsourced research project. In the Galaxy Zoo project, members of the public classify data from a million galaxies and publish the results in scientific journals. Facebook itself uses crowdsourcing to translate its pages into foreign languages.

Social networks also provide new ways for people to interact with government. The White House asked its Twitter followers for comments on a tax law.[5] An official from the National Economic Council then posted a blog with links to questions raised by the Twitter followers, eliciting a discussion about the direction tax policy should take. In 2011, the social network created by the city of San Francisco introduced a phone app that allowed citizens to take photos of potholes and other things that needed maintenance and upload them directly to the proper city office to order repairs. Through that same network, people with CPR skills can volunteer to help in an emergency. If someone has a heart attack on a golf course, a smartphone app will recruit volunteers in the area based on their GPS position and ask them to rush over to Hole 7 to render aid.

And when people get fed up with their government, they can use Facebook, Twitter, and YouTube to incite others to join them in the streets to protest. While previous forms of political protest required a charismatic leader, that leader could be killed or his headquarters destroyed. It's much more difficult to stop a widely dispersed group of antagonists such as the citizens of Facebook Nation. It's harder to put out thousands of revolutionary fires burning across the Web.

Social networks have enormous benefits, helping us stay in touch with people from our pasts and introducing us to people who share our interests. They create a much-needed comfort zone. As philosopher Ian Bogost points out, "Public spaces in general have been destroyed, privatized, and policed in recent decades, but the public life of teens and young adults has been particularly damaged, due to additional fears of abduction, abuse, criminality, and moral corruption."[6] According to Bogost, social networks provide a place to hang out, akin to the main drag or the video arcade of the past.

Social networks have become ubiquitous, necessary, and addictive. Social networking is no longer just a pastime; it's a way of life. People expect to be able to log on to Facebook or Myspace wherever they go and to tweet their every thought. Until recently, cell phones and internet use were banned in certain places, like courthouses, but now social institutions have largely abandoned their efforts to keep someone away from their Facebook friends or Twitter audience. As a result, there's a whole new set of issues, with judges friending defendants, jurors looking up witnesses' Facebook pages to assess their credibility, and lawyers blogging about confidential interchanges with their clients.

The military held out for a long time. In August 2009, the U.S. Marine Corps formalized its ban on marines' use of Myspace, Facebook, and YouTube on its networks.[7] The military's concern was the same as it is with many of us—phishing, hacking, and other security breaches. But the stakes were much higher. It's a hassle when you have to get a new credit card because your American Express number is hacked through PlayStation.[8] But it's much more serious if military design secrets are stolen by other countries or soldiers die when confidential battle plans are revealed.[9]

The military ban made sense except for two things. It was hard enough to get people to enlist in an all-volunteer armed services. But morale sank even lower when they were cut off from Facebook friends and Myspace family members. And the technology of armed conflict itself was demanding a link to the Web. For certain weapons to be used most effectively, soldiers need access to smartphone apps—such as iSnipe and Shooter—to estimate bullet trajectories. Another app allows soldiers to see the positions of friendly soldiers and enemy combatants on a map updated in real time.[10] There's even an app—Jibbigo—to translate a particular Iraqi dialect of Arabic.[11] And another, Telehealth Mood Tracker, to measure a soldier's mental health.[12]

In February 2010, the U.S. military embraced social networks in a big way. The military reconfigured its internet grid, NIPRNET (Non-classified Internet Protocol Router Network)—the largest private network in the world—to provide soldiers access to YouTube, Facebook, Myspace, Twitter, and Google apps.[13] The army began issuing smartphones to soldiers to test the apps' effectiveness both in and out of combat.[14] In war zones, wireless networks on which to run the apps are brought into the field attached to vehicles, planes, or air balloons.[15]

Not just our soldiers but our global enemies are taking to social networks. A 2010 Department of Homeland Security Report entitled "Terrorist Use of Social Networking Sites: Facebook Case Study" found that jihad supporters and mujahedeen are increasingly using Facebook to propagate operational information, including improvised explosive device (IED) recipes in Arabic, English, Indonesian, Urdu, and other languages.[16]

A 2,000-member militant Islamic Facebook group includes informational videos on "tactical shooting," "getting to know your AK-47," "how to field strip an AK-47," and so forth.[17] Facebook pages for other extremist Islamist groups contain propaganda videos featuring wounded and dead Palestinians in Gaza, links to Al Qaeda YouTube videos, and videos promoting female suicide bombers, all of which can be accessed by the public without becoming a "fan" of the groups, "liking" the groups, or "friending" the Facebook pages.

Even criminals use the Web for everything from figuring out who to rob by checking Facebook posts containing the word "vacation" to using a search engine to train for murder. Sometimes virtually the whole crime can be reconstructed from a search history, as in the case of a nurse who killed her husband after Googling "undetectable poisons," "state gun laws," "instant poison," "gun laws in Pennsylvania," "toxic insulin levels," . . . "how to commit murder," "how to purchase hunting rifles in NJ,". . . "neuromuscular blocking agents," . . . "chloral hydrate," "chloral and side effects," and "Walgreens."[18]

Facebook even facilitates real-time broadcasts of crimes in progress—and allows criminals to seek aid from friends. When Utah police tried to serve a warrant on Jason Valdez, a member of the Norteños gang, he barricaded himself in a motel room, taking Veronica Jensen as a hostage. With SWAT officers outside his motel room and in the adjoining rooms, Jason used his Android phone to post six status updates to Facebook, add 15 friends, respond to numerous comments on his wall posted by friends and family, and post a picture of himself and his hostage with the caption, "Got a cute 'HOSTAGE', huh?" A Facebook friend posted on Jason's page that a SWAT officer was hiding in the bushes: "gun ner in the bushes stay low." "Thank you homie," Jason replied. "Good looking out."[19]

Eight hours later, Jason posted his last update: "Well I was lettin this girl go but these dumb bastards made an attempt to come in after I told them not to, so I popped off a couple more shots and now were startin all over again it seems." The standoff ended when SWAT officers used explosives to blast through the front door and through the wall from an adjoining room. The hostage was fine, Jason ended up in intensive care, and police are considering whether to charge Jason's friend with obstruction of justice for warning him about the SWAT officer.[20]

It's easy to understand why people flock to Facebook and other social networks. But it's harder to anticipate what will happen to you when you become a citizen of this new world. If you were to move to a kibbutz in Israel, teach English in Japan, enlist in the army, or move to a rural farm, you'd have some sense of what you were getting into. When you join Facebook, you don't know enough about the ramifications of social network citizenship to understand where that decision will lead and how it might transform you and your life. The governing rules of Facebook—its terms of service—shift rapidly and without warning. One day, it promises you that your friends are private; the next day, it makes them public.

One might think that Facebook enhances the Constitutionally protected freedom of association since it allows groups to form. Class of 1995 Reunion Committee. I Love Justin Bieber :)). Free the West Memphis Three. Yet people's Facebook associations have been used against them. Judges have been disciplined for "friending" lawyers on Facebook, even though it is completely acceptable for judges to be friends with attorneys in real life. In one case, a prison guard in England was fired after he friended prisoners on Facebook.[21]

And a crucial part of the freedom of association is the right not to reveal your associations. In 1958, for example, the U.S. Supreme Court allowed the civil rights organization for African Americans, the NAACP, to keep its member list secret from the government of Alabama. The NAACP argued that compelled disclosure would "abridge the rights of its rank-and-file members to engage in lawful association in support of their common beliefs."[22] In deciding not to compel disclosure, the Court held that people might be afraid to exercise their freedom of expression and engage in collective action to further those beliefs if their membership in the organization was known. Exposing a person's membership in a group could lead to "economic reprisal, loss of employment, threat of physical coercion, and other manifestations of public hostility."[23]

Yet social networks have revealed associations that people had expected to be private. When, without any notice, Facebook changed its policy in 2009 so that lists of friends and affiliations were made public and no longer subject to people's privacy controls, the repercussions were felt around the world. In Iran, authorities questioned or detained the Facebook "friends" of Tehran's U.S.-based critics. People were beaten. Americans traveling to Iran were detained and had their passports confiscated just for having a Facebook page.

Unlike in a democracy, Facebook is unilaterally redefining the social contract—making the private now public and making the public now private. Private information about people is readily available to third parties. At the same time, public institutions, such as the police, use social networks to privately undertake activities that previously would have been subject to public oversight. Even though cops can't enter a home without a warrant, they scrutinize Facebook photos of parties held at high school students' homes. If they see the infamous red plastic cups suggesting that kids are drinking, they prosecute the parents for furnishing alcohol to minors.

You might think you are posting information just to family members, but with a modest change in computer code, the privately run Facebook can send that information anywhere. Both inadvertently and through conscious decisions, Facebook and other social networks have put private information, including medical test results, credit card numbers, and sensitive photos into the wrong hands.

Unlike Vegas, what happens in Facebook doesn't stay in Facebook.

The precursors to social networks were multiplayer online games, where people interacted through avatars they'd created that often bore no resemblance to the real person they represented. But what you do in today's social networks doesn't involve playing a role. And the implications are much greater than winning and losing loot, treasures, and experience points.

Unlike games and previous social networks, Facebook asks the user to submit his or her real name and email address. And the actions taken on Facebook and other social networks have real world consequences. Women have been fired because their Facebook posts showed them wearing provocative clothing. Straight-A students have been expelled from school for criticizing their teachers on Myspace. When, the day before a criminal trial, a cop posted on Myspace that his mood was "devious," a parolee charged with gun possession used that post to persuade a jury that the cop had planted a gun on him.[24]

Colleges and companies routinely search Facebook and Myspace to determine whether to admit or to hire people. A background-checking service called Social Intelligence Corp. accumulates files on the Facebook photos and posts of any user with privacy settings marked "Everyone."[25] The company keeps each person's file for seven years—so even if you delete that photo of you in your "Free Charlie Manson" shirt, you're still going to have a hard time getting a job.

Artists have demonstrated how information posted on Facebook can be used out of context. Italian artists Paolo Cirio and Alessandro Ludovico created Lovely Faces.com, a fake dating website using information from publicly available Facebook profiles.[26] They used software to copy people's names, pictures, and locations from over a million Facebook profiles.[27] Their software extracted 250,000 faces, and the artists used a facial recognition algorithm assessing features and expressions to classify the photos into six categories—social climber, easygoing, funny, mild, sly, and smug.[28] Paolo and Alessandro discussed the impetus behind their project when it appeared at a Berlin art festival: "Being judged is the price everyone must pay to be involved in social networks. The project sneers at the trust that is brought to the platform by 500 million users, reminding them that there are—just as in the 'real' world—consequences to publishing intimate, personal information on social networks."[29]

Our digital identities on the Web—email, personal websites, and social media pages—are starting to overshadow our physical identities. As we work and chat and date (and sometimes even have sex) over the Web, we are creating a digital profile of ourselves that redefines us—and could come back to haunt us.

Not only does Facebook make the private public, it makes the public private. Government officials used to have to obtain warrants to find out private facts about people. Now they can monitor Facebook postings and Google searches to gain access to intimate and revealing information about people. Law enforcement officials troll public profiles for clues to crimes or to anticipate emergency situations, like the Department of Homeland Security's monitoring of the use of certain terms on social networking sites—terms ranging from "Organized Crime" to "Quarantine."[30] In its January 2011 initiative, the Department of Homeland Security listed 350 terms to be monitored, including "Pork," "Vaccine," "Pirates," "Body Scanner," "Guzman," and even "Social Media."

A 2008 U.S. Citizenship and Immigration Services memo even recommended friending citizenship petitioners to monitor the validity of their relationships, searching for evidence, for example, that a marriage doesn't meet the department's legitimacy standards.[31] Noting that the "narcissistic tendencies in many people fuels a need to have a large group of 'friends' . . . provides an excellent vantage point for FDNS [Office of Fraud Detection and National Security] to observe the daily life of beneficiaries and petitioners who are suspected of fraudulent activities," the memo also identified social networking sites popular among certain ethnic groups (such as MiGente.com, popular with Latinos, and Muxlim.com, popular among Muslims).[32]

Governments have even started crowdsourcing investigations. With ever-present cell phone cameras—and widely distributed computing power, citizens can go beyond watching cop shows to working cases themselves. When a scandal erupted in England about the expense reports of Members of Parliament, more than 450,000 pages of expense accounts were posted online. More than 28,000 people joined in the digital search for irregularities in the MPs' spending.[33]

The State of Texas installed 29 surveillance cameras along the Texas-Mexico border (one camera for every 41 miles) and allowed anyone with an internet connection to monitor the border and alert the authorities about alleged illegal immigrants and drug traffickers.[34] The website, powered by the social network BlueServo, allows people from all over the world to become "Virtual Texas Deputies."[35] On the website, instructions appear next to each camera's feed, such as "Report anyone crawling through this culvert" or "Look for subjects on foot moving towards right."[36] A virtual deputy in Australia watches the border from a pub;[37] a mom in New York spends four hours a day fixated on the desert images while caring for her infant daughter.[38] And an Oklahoma woman, who visits BlueServo each night after work and also tracks bald eagles online, describes her involvement, "I watch eagles and illegals. That's a fun thing to do."[39]

Launched in November 2008 with a $2 million grant from the Texas governor's office, the site attracted 130,000 virtual deputies in its first year.[40] But a year later, and despite another $2 million grant from the governor, the project had resulted in a total of only 26 arrests—a cost of $153,800 per arrest—far short of the 1,200 arrests the State had projected.[41] The project floods sheriffs' offices with thousands of emails each week, describing an "armadillo by the water" or falsely reporting, "There are some men crossing the water. They have a bottle of tequila and a big hat with them."[42]

Although supporters describe the virtual deputies as an "extra pair of eyes" for law enforcement, behind those eyes may be strong personal sentiments about the dangers of illegal immigration. As Texas State Senator Eliot Shapleigh observed, the cameras "invite extremists to participate in a virtual immigrant hunt."[43] Indeed, reports submitted by virtual deputies are often phrased in racist terms, such as "Two wetbacks and a dog walking across screen."[44] Critics of the project worry that the 24/7 surveillance cameras will encourage vigilantism—instead of reporting perceived illegal immigration, virtual deputies may "jump in their truck with a gun," notes Jay Stanley of the American Civil Liberties Union.[45]

And BlueServo live streams images not only from the Texas-Mexico border but also from the streets of average American neighborhoods. The site allows people to connect their own cameras to the network to patrol their local neighborhoods.[46]

Unlike the deserted lands along the border, trafficked largely by animals, neighborhood cameras may capture people engaged in a variety of routine activities, from teenagers kissing in a car to mothers spanking their children in the driveway. Because virtual deputies are not committed to the principles of just law enforcement, they are more likely to engage in selective reporting of incidences—getting back at a neighbor who plays loud music or targeting a member of a disliked population.

Some social networks, like 4chan (one of the internet's most trafficked sites), have used the digital equivalent of deadly force to police society. When 4chan users discovered a video in which a teenager seemed to be abusing a cat, they vowed to track him down and have him arrested. The 4chan users watched other videos on the boy's YouTube account and matched the background room details in one to a photo found on a social network site.[47] The online manhunt continued to search for the boy on various social networks and finally found his Facebook page.[48] They traced the boy to Lawton, Oklahoma, and reported him to the local police.

In other instances, 4chan members themselves have "punished" someone they think is an offender—by publishing the target's address, hacking into his computer, and overwhelming him with crank calls and emails. When Sarah Palin was accused of violating campaign finance laws by using her gubernatorial email account to solicit funds, a college student hacked into Palin's email account, published her password on the 4chan website, and circulated screenshots of the emails she was sending.[49]

Social networks are transforming the activities of cops and judges, often challenging cherished principles of democracy. Police sweep social networks for signs of misconduct and create fake profiles to befriend and monitor gang members.[50] The IRS searches Facebook and Myspace profiles for evidence of taxable transactions and the whereabouts of tax evaders.[51] Courts consider social network posts about a parent's partying to be signs of parental neglect. Although such institutional uses may be touted as furthering law enforcement, they conflict with traditional due process rights and the principle that citizens should be free from constant scrutiny.

These trends raise troubling questions. Governments in democracies exist to protect their citizens and uphold their rights. Public health officials, like other government agents, are limited in the type of data they collect and publicize about citizens. Similarly, the police need probable cause and search warrants to collect evidence. Constitutional restrictions on governmental action apply. But what happens when state and federal agencies circumvent those rules by turning the job of evidence gathering over to private citizens using social networks—or ask the social networks themselves for private information from people's photos and posts? Should that evidence also be subject to Constitutional constraints?

Google searches involving flu symptoms provide a better indicator of the geographic areas where a flu outbreak is spreading than do traditional public health measures.[52] But should state and federal public health officials be allowed access to private searches? And what if they learn that someone is searching for information about a communicable disease? Should public health officials be able to track down that person and quarantine him? When citizen pseudo-cops reach the offender before the real cops do, what's to prevent them from taking the law into their own hands? As the line between governments and social networks blurs, who should be held accountable when something goes wrong?

Facebook does more than just supplement governmental functions. Facebook is a nation in its own right. It has an economy. It has certain rules of governance. But unlike any democratic nation, its operating procedures are more like a computer manual than a Constitution. All the rights run in one direction. Facebook holds the cards, and its citizens have little recourse—other than to leave the service entirely. In fact, several federal laws—created for the internet before social networks were even contemplated—protect social networks from almost any liability. Because of these laws, social networks cannot be sued for invasion of privacy, defamation, or criminal acts based on people's postings.

Social networks turn people's private information into the networks' own income streams. Facebook makes most of its money as an advertising platform.[53] Facebook sells ad space on its site and helps advertisers personalize their advertisements and direct them toward specific members by using the information from the Facebook member's profile and entries.[54] Advertisers choose keywords or details—such as relationship status, location, interests, activities, favorite books, demographics, employment information—and then Facebook runs the ads for the targeted users.[55] eMarketer estimated that Facebook earned $1.86 billion in advertising revenue in 2010.[56]

Facebook also makes money through revenue-sharing agreements with companies that offer the more than 550,000 applications that run on the site,[57] including games such as Mafia Wars and FarmVille. In 2011, Facebook told all game developers that they will be required to accept payments through the social network's new currency, Facebook Credits.[58] This will have a significant impact on Facebook's earnings because Facebook will keep 30% of the amount of the credits, gaining a sizeable chunk of the half-billion dollars earned annually by apps.[59]

Facebook is not the only company that capitalizes on the intimate information people post online. While a student at Stanford, Harrison Tang and his friends created Spokeo, a search technology that pulls together information about millions of identifiable individuals from social networks and far-flung areas of the Web.[60] Spokeo not only compiles information from hundreds of online and offline sources—ranging from real estate listings to marketing surveys[61]—but it also uses that information to generate characterizations about individuals, such as "self-driven," "donates to causes," "has veterans in the house," and "collects sports memorabilia." By entering a person's name on the Spokeo website, you can view the person's address, home phone number (even if unlisted), age group, gender, ethnicity, religion, political party, marital status, family members, and education for free.[62] Spokeo often includes a Google Maps image of the person's residence.

Spokeo bills itself as "not your grandma's white pages."[63] For less than five dollars a month,[64] you can view more extensive information, including attributes of the person's property (such as whether the property has a swimming pool or fireplace), lifestyle and interest information, and Spokeo's assessment of a person's wealth level (such as "bottom 50%") and economic health (such as "average" "or "very strong"). The five-dollar-a-month subscription also permits a limited number of reverse searches using a person's email address or user name.[65] These searches can retrieve a person's profiles on social networking sites and dating sites, playlist on Pandora, photos posted on sites like Flickr or PhotoBucket, videos posted on YouTube, blogs, and reviews of products on shopping sites like eBay,[66] if those posts have not been protected by privacy settings. Spokeo's "Enterprise" subscription, which costs $79.95 a month, allows the subscriber to search up to 1,000 email addresses and 1,000 user names each month.[67]

More than a million people search the Spokeo database each day[68]—making decisions about whether to hire people, grant them credit, or even have sex with them—based on what they read. Often the information is wrong, culled from erroneous or outdated sources and interpreted by imperfect algorithms. Yet the people whose privacy is being invaded may not even realize that Spokeo exists, let alone that it has stigmatized them. Spokeo does not believe that it is subject to laws that regulate credit reporting bureaus by requiring them to remove or correct inaccurate or unverified information and limiting who can see the information. The Act applies to entities that collect information about people's creditworthiness, personal characteristics, mode of living, that could be used to determine eligibility for credit or employment.[69] Spokeo claims that it is "intended for entertainment purposes only" and that it should not be "considered for purposes of determining a consumer's eligibility for credit, insurance, or employment."[70] But Spokeo promotes its service as providing "invaluable insight" into the people whose data Spokeo offers and has featured banners on its site that said "HR Recruiters— Click Here Now!" and "Want to See Your Candidates' Profiles on MySpace and LinkedIn?"[71] Spokeo has also advertised on other websites, including through an ad that said, "Is he cheating on you? Reverse search his email to find out."

When I told a law school professor about Spokeo, he logged on to read what assumptions they'd made about his life. Spokeo correctly identified his home and listed accurate home and cell phone numbers. Because his wife's name was Jamie, Spokeo had assumed she was his young son. And Spokeo had knocked 30 years off his age and assumed he was married to his 30-year-old daughter, who had, of course, the same last name. The mistakes could affect his credit rating, since a 30-year-old would be assumed to earn less than a 60-year-old. But how would you even know to go on Spokeo to try to correct what was there? My colleague taught computer law and he hadn't even heard of Spokeo before I mentioned it.

When Thomas Robins looked himself up on Spokeo, he noticed that the site had lots of things wrong. Spokeo had incorrectly stated that Robins was in his 50s, was married, was employed in a professional/technical field, had a graduate degree, and had children. Spokeo also provided a photo of a person whom it misidentified as Robins and, according to Robins, inaccurately reported his "wealth level." Robins was concerned that the inaccuracies in his report would affect his ability to obtain credit, employment, insurance, and the like.[72] He sued Spokeo. The judge dismissed his claim under California's Unfair Competition Law because Robins hadn't alleged that he'd suffered an actual harm as a result of Spokeo's actions.[73] The judge allowed the suit to go forward under the federal Fair Credit Reporting Act, though. According to the court, "Plaintiff's allegations that Defendant regularly accepts money in exchange for reports that 'contain data and evaluations regarding consumers' economic wealth and creditworthiness' . . . are sufficient to support a plausible inference that Defendant's conduct falls within the scope of the [Fair Credit Reporting Act]."[74] But the judge later changed his mind on the Fair Credit Reporting Act claim, stating that the alleged harm to Robins's employment prospects is "speculative, attenuated, and implausible." If a claim like Robins's under the Fair Credit Reporting Act were allowed to go forward, "courts will be inundated by websurfers' endless complaints," the judge said. Robins has appealed the decision to a higher court.

Spokeo is part of a sprawling multibillion-dollar industry of data aggregators[75] — organizations that collect information from various data repositories, such as public records, criminal databases, and social network sites.[76] The information is packaged into reports and sold to other organizations, such as advertising firms, businesses, government agencies, and credit card companies.[77] The data reports can be used for employee background checks, marketing research purposes, security, creating targeted mailing lists, or determining which ads appear on social networks and other websites.[78]

Some marketing companies have even made deals with internet service providers to tap into any information people send from their computer, be it a private post to Facebook, an email to a lover, or a Google search for "how to commit murder." Amazingly, courts have said that such commercial arrangements don't violate the wiretap laws. According to one judge, the right to privacy "is lost, upon your affirmative keystroke."[79]

It's always a tip-off to something wrong in a nation when its leaders live by different rules than those governing citizens. Although Harrison Tang created Spokeo to collect and sell people's personal information, he decided to remove himself from the database: "I was getting a lot of emails and threats," he stated.[80] Yet Spokeo provides home addresses, unlisted phone numbers, and private information of other people without their permission, despite the fact that the availability of the information could lead to threats to those other people.

And even though Facebook monetizes people's private information in deals with advertisers and game designers, Facebook booted off one of its users when he tried using a computer program to copy his own data — his friends list — off his Facebook page.[81] Facebook also threatened legal action against the artists behind the Lovely-Faces project and terminated the artists' personal Facebook accounts.[82] The artists argued that their "conceptual art provocation" used publicly available information and was therefore legal.[83] But the artists buckled under pressure from

Facebook's lawyers. They took down Lovely-Faces.com, but continue to host Face-to-Facebook.net, which explains the project, including their legal travails.[84]

People are understandably drawn to social networks. For individuals, social networks allow people to stay in touch, performing some of the same functions performed by telephones and letters in previous eras. But laws protect us against outsiders tapping our phones and reading our private mail. Even prisoners can send mail to their lawyers without having those letters read by prison officials. But everything we post on social networks is fair game for the engineers behind Facebook and any other data miner.

Facebook and other social networks are transforming huge swaths of our lives—how we mate, shop, work, and stay in touch with the people we love. They are also changing the political process itself. When John F. Kennedy and Richard Nixon debated on television, concerns were raised that politics would deteriorate into a contest where the most telegenic candidate won. But TV debates took place out in the open—anyone could tune in. And the Federal Communications Commission adopted regulations so that opposing candidates were granted equal time to present their views.

With social networks, it's not the most telegenic candidate who wins, but the one with the best data crunchers. Barack Obama was swept into office largely because of his presence on the Web.[85] His social network campaign was managed by one of the founders of Facebook, 24-year-old Chris Hughes, who took a leave from the company to propel Obama into office.

The Republicans did Obama one better and stormed Washington in the 2010 elections through targeted use of social network data. Data aggregators used data from social networks, such as people's interest in the Bible, past political contributions, voter registration status, shopping history, and real estate records to identify conservative voters by name and provide that information to Republican political hopefuls. The candidates could then email the people directly, making promises and taking stances that were never revealed to the public—and were shielded from scrutiny by their opponents.

With not only the rights of individuals at stake, but the future of the political process itself, it's time to analyze how we as social network citizens can be protected. What responsibilities should individuals bear? What rules should govern what can be done with our digital selves and our data by the social networks themselves and the third parties who gain access to that information? What rights should social network citizens have?

The complex issues raised by social networks came to the fore after the 2011 British riots. Prime Minister David Cameron, who'd previously felt that social network communities were "brilliant," felt differently once rioters began to communicate with each other via Facebook, Twitter, and BlackBerry Messenger to share information about what shops to loot.

"Everyone watching these horrific actions will be struck by how they were organised by social media," the prime minister told the House of Commons. "So we are working with the police, the intelligence services and industry to look at whether it would be right to stop people communicating via these websites and services when we know they are plotting violence, disorder and criminality. I have also asked the police if they need any other new powers."[86]

Member of Parliament David Lammy pointed out that rioters had used Black-Berry Messenger to send encrypted and almost untraceable messages to each other. He urged Research in Motion, the maker of BlackBerry, to shut off that service entirely until order was restored in the streets.[87] The prime minister similarly asked Twitter and Facebook to remove messages, images, and videos that could incite riots.[88]

Civil rights advocates reacted immediately. "How do people 'know' when some-one is planning to riot?" asked Jim Killock, the executive director of online advo-cacy organization Open Rights Group. "Who makes that judgment?"[89] Legitimate advocacy and well-grounded protests will be stifled if social networks and websites are pressured to censor their members.

Social networks have stunning benefits. But the citizens of Facebook Na-tion who see those benefits may not realize the downside. The young nation was founded only recently, less than a decade ago. Its original citizens were college stu-dents who are probably still too young to have experienced rampant discrimination in jobs, romance, or credit lines based on what they've posted. They may not yet realize the extent to which their offline self is being overshadowed by their digital doppelgänger.

People came to Facebook Nation for freedom of association, free expression, and the chance to present an evolving self. But unless people's rights are protected, social networks will serve to narrow people's behavior and limit their opportunities, rather than expand them. Already people are being fired for undertaking perfectly legal activities, such as when a photo of an employee drinking wine is tagged on Facebook. And new norms of behavior are emerging that do not reflect off-the-grid life, such as forbidding judges to "friend" lawyers.

Social networks are taking over many of the traditional functions of government without any legal protections for their citizens. The underlying economic goal of social networks—monetizing personal data—is invisible to their citizens and may in fact be herding them into a land that they wouldn't want to inhabit.

The United States Constitution was penned by philosopher-politicians gravely concerned with the metaphysical question of what was necessary for individual and social flourishing. They understood that living socially and with aspirations meant adopting principles to deal with everything from resolving disputes to encouraging innovation, from structuring relationships with other nations to protecting indi-vidual rights.

They recognized the value of protecting people's privacy and assuring oversight of actions of the government. They required that the governing rules about the relationship between citizens and the government be clearly stated in advance and not changed without adequate notice and citizens' input. They favored openness about what the government was doing, believing, as U.S. Supreme Court Justice Louis Brandeis said a century later, "sunshine is the best disinfectant." They also saw the value of being able to remake oneself, to start afresh.

Instead of philosophy, computer engineering and data collection are the driving forces behind the policies of Facebook Nation. The quest for more and more information about more and more people is what stimulates the Facebook economy because the service makes its money on data. The executives behind social networks often disregard the values that are central to the U.S. Constitution. The Facebook founders, for example, view the desire for privacy as something to be outgrown. In a 2010 interview, Mark Zuckerberg commented on Facebook's decision to make certain previously private information public: "People have really gotten comfortable not only sharing more information and different kinds, but more openly and with more people."[90] Former Facebook programmer Charlie Cheever said, "I feel Mark doesn't believe in privacy that much, or at least believes in privacy as a stepping-stone."[91]

And the very structure of social networks prevents you from reinventing yourself. Once information about you and photos of you are on the Web, they can be used against you in perpetuity.

As we each begin to live a parallel life on Facebook, it's time to figure out, as with any new country, what principles should govern this new nation. Do the principles on which the United States and other democracies were founded still resound with people today? Could they provide guidance for the governance of social networks?

The project of proposing a Social Network Constitution may seem foolish. Facebook, Myspace, Google, Twitter, and YouTube are private entities, and the U.S. Constitution governs only the actions of the government, not private actors. But that is not the case in other countries, such as Germany, Ireland, South Africa, and the European Union,[92] where the fundamental values expressed in the national constitution can apply to companies in addition to governments. After all, companies may be more powerful than some governments—that's certainly the case with Facebook.

And even in the United States, the fundamental values expressed in the U.S. Constitution provide guidance for the private realm. The Fourteenth Amendment's idea of equal protection under the law provided the foundation for Congress to enact civil rights laws that govern the conduct of corporations and private citizens. The Fourth Amendment's protections for privacy provided judges with the inspiration to allow lawsuits against individuals and corporations that disseminated a person's private information without consent.

We needn't think of a Social Network Constitution as a set of rules, like the Internal Revenue Code, that would govern in minute detail what a social network should or shouldn't do. Instead, think of it as a touchstone, an expression of fundamental values, that we should use to judge the activities of social networks and their citizens. These principles could be used to frame the societal debates about social networks—guiding not only the decisions of citizens about what technologies they should reject but also the decisions of courts and legislatures about what principles should govern.

In many instances, the principles would help courts make a determination in a case, analyze existing laws, and decide whether or not to let evidence in at trial. These values could also guide legislators who are considering adopting new laws to regulate social networks.

The very nature of social networks is constantly changing. New technologies are introduced and individual users face new issues. A set of strict, rigid rules governing the use of social networks might be effective now, but will quickly become outdated just as other laws that are intended to protect people, such as wiretapping laws and consent laws, fail to protect and serve the needs of the current online community. Unlike the rigid, formula-driven Internal Revenue Code, a Social Network Constitution should be flexible and recognize basic principles that we should never outgrow. Its provisions would address the actions of government agencies, social institutions, and society at large.

Every democratic nation has governing principles about what rights its citizens have over property, privacy, life, and liberty. The citizens of Facebook Nation deserve no less.

2

George Orwell . . .
Meet Mark Zuckerberg

On a Sunday morning, I fire up my laptop and compose a memo to my co-counsel about a pro bono case we are considering filing against a biotechnology company. I attach it to an email and send it to him, carefully writing, "Confidential—Legal Mail" in the subject line and putting a few key ideas in the text of the email. Then I log on to the Southwest Airlines site, enter my credit card information, and buy a ticket for Florida. I enter a governmental website, run by the Florida Fish and Wildlife Conservation Commission, and type in my Social Security number to obtain a fishing license. I realize I'll be away on my sister's birthday and send her some books from Amazon. I check my emails and click through to a website that lists job openings for university professors. One is in a town I haven't heard of, so I Google it to find out if it will be urban enough for me. The town's name brings up a link to a local newspaper article about a poisoning and I save that information to my hard drive, thinking I might use it in the next mystery book I write. I read an email from my doctor telling me she changed my prescription electronically and the new drug is waiting for me at my neighborhood CVS. Before leaving the house to pick it up, I log on to Facebook to contact friends in Florida and let them know when I'll arrive. Elsewhere on my Facebook page, I check my news feed and indicate I liked the movie I saw the previous night. Someone has tagged me in a Halloween photo from years ago, when I was a Yale undergrad. I am wearing a belly dancer's costume and I am with someone dressed like a bottle of Imperial Single Malt Scotch. I untag myself from the photo. If I do interview for a new job, I don't want someone to say to me, "Well, Ruth Bader Ginsburg would never have shown her navel."

All in all, I feel good about the security of my morning's travels across the Web. I haven't responded to any wealthy widows seeking my legal help for their $50 million estates, nor to emails purportedly from friends whose wallets and passports were stolen in London. I haven't given my credit card to anyone with a sketchy foreign email address who offers me an iPad for $30, nor have I opened the missive that tells me I've exceeded my email limit. I've only dealt with websites I trust.

But every action I've taken has been surreptitiously chronicled and analyzed by data aggregators, who then sell the information to companies, including perhaps the one I am contemplating suing. And not only have I not been informed about this invasion of my privacy and security, there's almost nothing I can do about it.

That stunning fact is completely at odds with the offline world. I care deeply about the type of information I've entered. I wouldn't leave my Social Security number or my credit card number lying on my desk at work where someone could copy it—nor would I send that information on a postcard through the mail. I wouldn't broadcast my medical condition or my desire to find a new job to the world. But that information about me is bought and sold daily by corporations that deal with data aggregators.

If someone broke into my home and copied my documents, he'd be guilty of trespass and invasion of privacy. If the cops wanted to wiretap my conversation, they'd need a warrant. But without our knowledge or consent, virtually every entry we make on a social network or other website is surreptitiously being tracked and assessed. The information is just as sensitive. The harms are just as real. But the law is not as protective.

The guiding force behind this enormous theft of private information is behavioral advertising. The covert collection of personal information is an exploding industry, fueled in part by the lust of advertisers for personal data about people's habits and desires. "Online behavioral advertising," notes the Federal Trade Commission, "involves the tracking of consumers' online activities in order to deliver tailored advertising. The practice, which is typically invisible to consumers, allows businesses to align their ads more closely to the inferred interests of their audience."[1] But the unregulated amassing of personal information about people has also been used in ways that cause them harm.

Behavioral advertising was used by 85% of ad agencies in 2010.[2] They're drawn to it because it works—63% of ad agencies say targeted ads increased their revenue, with 30% of agencies reporting that behavioral advertising increased their revenue by $500,000 or more. In 2010, internet advertising revenues exceeded that of newspapers by $3.2 billion.[3] During the first quarter of 2010, internet users in the United States received 1.1 trillion display ads, which cost the ad sponsors about $2.7 billion.[4]

"It's a digital data vacuum cleaner on steroids, that's what the online ad industry has created," Jeff Chester, executive director of the Center for Digital Democracy, told *The New York Times*. "They're tracking where your mouse is on the page, what you put in your shopping cart, what you don't buy. A very sophisticated commercial surveillance system has been put in place."[5]

It's through data aggregation that Facebook makes its money. Facebook sits on a mountain of information worth a fortune. It's expected that in 2012 Facebook will be valued at $100 billion.[6] Currently the company generates most of its revenue by acting as an intermediary between advertisers and its database of users' personal information. Facebook will use information about my status, likes and dislikes, and the recent post about my travel plans to update its digital portrait of me. When an airline or outdoor clothing company pays Facebook to post an ad for traveling adults, Facebook will use its new information about me to post the ad on my Facebook page. This commercialization of my private data—the information I think I'm only posting to friends—is the reason Facebook earned an estimated $1.86 billion in 2010 from the display ads, 90% of its total revenue, and was expected to bring in $4.05 billion in advertising revenue the following year.[7]

Facebook uses its citizens' demographic information, interests, likes, friends, websites frequented, and even contact information as the foundation of its advertising platform. Facebook encourages users to disclose more information about themselves through "very powerful game-like mechanisms to reward disclosure," said media activist Cory Doctorow, co-editor of Boing Boing.[8] Doctorow compares Facebook's mechanisms to the famous Skinner box used in psychology experiments.[9] But instead of a lab rat rewarded with a food pellet each time it pushes a lever in the box, a Facebook user is rewarded with "likes" and attention from friends and family each time that person posts more information.

"And this is not there because Facebook thinks that disclosing information is good for you necessarily," says Doctorow. "It's in service to a business model that cashes in on the precious material of our social lives and trades it for pennies."

But the collection and marketing of personal information are far more insidious, and profitable, than just the actions of Facebook. Mark Zuckerberg's brainchild makes up only 14.6% of the behavioral advertising market. And some of the other advertisers use tactics that make Zuckerberg's seem tame. Every single action I undertook that Sunday morning was potentially seized by a data aggregator through some means or another. In California, consumers are suing the company NebuAd, which contracted with 26 internet service providers, including Delaware's Cable One, New York's Bresnan Communications, and Texas's CenturyTel, to install NebuAd's hardware on those internet service providers' networks without ISP users' consent.[10] The hardware allowed NebuAd to use deep packet inspection—a mechanism to intercept and copy all the online transmissions of the ISPs' subscribers and transmit them to NebuAd's headquarters.[11] All of them.

Everything you post on a social network or other website is being digested, analyzed, and monetized. In essence, a second self—a virtual interpretation of you—is being created from the detritus of your life that exists on the Web. Increasingly, key decisions about you are based on that distorted image of you. Whether you get a mortgage, a kidney, a lover, or a job may be determined by your digital alter ego rather than by you.

In the late 1960s, sociologist John McKnight, then Director of the Midwest Office of the U.S. Commission on Civil Rights,[12] coined the term "redlining" to describe the failure of banks, supermarkets, insurers, and other institutions to offer their services in inner city neighborhoods.[13] The term came from the practice of banks, which drew a red line on a map to indicate where they wouldn't invest.[14] But use of the term expanded to cover a wide array of racially discriminatory practices in general, such as not offering home loans to African Americans, even if they were wealthy or middle class.

Now the map used in redlining is not a geographic map but the map of your travels across the Web. A new term, "weblining," covers the practice of denying certain opportunities to people due to observations about their digital selves. Sometimes redlining and weblining overlap, such as when a website uses zip code information from a social network or an online purchase elsewhere to deny a person an opportunity or charge him a higher interest rate.

"There's an anti-democratic nuance to all of this," says New York University sociologist Marshall Blonsky. "If I am Weblined and judged to be of minimal value, I will never have the products and services channeled to me — or the economic opportunities — that flow to others over the Net."[15]

Data aggregation is big business. The behemoth in the industry, Acxiom, has details on everything from your Social Security number and finances to your online habits.[16] Its former CEO, John Meyer, described it as "the biggest company you've never heard of."[17] Rapleaf is another data aggregator that combines online data, including usernames and social networks, and offline data from public records.[18] One of its competitors, ChoicePoint, has acquired more than 70 smaller database companies and will sell clients one file that contains an individual's credit report, motor vehicle history, police files, property records, court records, birth and death certificates, and marriage and divorce decrees.[19] Yet ChoicePoint didn't do a great job of keeping that information secure. In 2005, identity thieves who submitted false applications to ChoicePoint claiming to be small businesses were given access to ChoicePoint's database that contained financial records of more than 163,000 consumers.[20] The Federal Trade Commission attributed the security breach to a lack of proper security and record handling procedures and, as part of a settlement with ChoicePoint, required the company to implement a comprehensive information security program and to pay $10 million in civil penalties and $5 million to reimburse the consumers affected by the identity theft.[21] That same year, hackers targeted LexisNexis (an aggregator which later bought ChoicePoint for $4.1 billion in cash), and accessed the personal information of 310,000 customers.[22]

Weblining goes further than traditional redlining. Sometimes an individual's credit card limit is lowered, midcourse, based on data from aggregators, even when the cardholder has done nothing wrong. Kevin Johnson, a condo owner and businessman, held an American Express card with a $10,800 limit. When he returned from his honeymoon, he found that the limit had been lowered to $3,800. The switch was not based on anything Kevin had done but on aggregate data. A letter from the company told him: "Other customers who have used their card at establishments where you recently shopped have a poor repayment history with American Express."[23]

Not only does weblining affect what opportunities are offered to you (in the form of advertisements, discounts, and credit lines), it also affects the types of information you see. When you open Yahoo! News or go to other news websites, you get a personalized set of articles, different from your spouse's or neighbor's. That may sound like a good thing, but you may be losing out on the big picture. With the physical version of *The New York Times*, you'd at least see the headlines about what was going on in the world, even if you were skimming the paper to get to the movie reviews. But world news may disappear entirely from your browser if you have indicated an interest in something else. Ever since I clicked on a story about the royal wedding, the world news stories I used to receive when I logged on to my email have been replaced by celebrity breakup and fashion stories. But if we are all reading a different, narrow range of articles, how can we participate in a civic democracy?

"Ultimately, democracy works only if we citizens are capable of thinking beyond our narrow self-interest. But to do so, we need a shared view of the world we cohabit," says Eli Pariser in *The Filter Bubble: What the Internet Is Hiding from You*. Pariser explains that the internet initially seemed like the perfect tool for democracy. But now, he points out, "Personalization has given us something very different: a public sphere sorted and manipulated by algorithms, fragmented by design, and hostile to dialogue."[24]

Most people have no idea how much information is collected surreptitiously about them from social networks and other websites. When asked about behavioral advertising, only half of the participants in a 2010 study believed that it was a common practice.[25] One respondent said, "Behavioral advertising sounds like something my paranoid friend would dream up, but not something that would ever really occur in real life."

People have a misplaced trust that what they post is private. A Consumer Reports poll found that "61% of Americans are confident that what they do online is private and not shared without their permission" and that "57% incorrectly believe that companies must identify themselves and indicate why they are collecting data and whether they intend to share it with other organizations."[26]

When people realize that websites and advertising companies are collecting extensive information about them, many want legal change. A telephone survey found that 66% of adult Americans opposed being targeted by behavioral advertising and are troubled by the technologies used to enable it.[27] Also, 68% of Americans opposed being "followed" on the Web and 70% of Americans supported the idea of requiring hefty fines to be paid by a company that collects or uses someone's information without his or her consent. Most people—92%—believe that websites and advertising companies should be required to delete all information stored about an individual if requested to do so.

Your ability to protect yourself against unwanted data collection depends largely on the technique being used to acquire information. With some methods, companies use your own computer against you by instructing your internet browser to store information on your computer's hard drive that data aggregators can use to track your movement online and build a profile of your online behaviors. Other methods tap the information as it travels from your computer to the recipient's website or email address. (See "Web Tracking Chart.")

The collection of information from websites and social networks began modestly enough. Social networks asked you if you'd like to have your password stored. Websites like Amazon.com began to keep track of your purchases on their sites to make recommendations and to allow you the convenience of not re-entering a password or credit card number each time you visit the site. Now tracking technologies with names like cookies, Flash cookies, web beacons, deep packet inspection, data scraping, and search queries allow advertisers to create a picture of you by noting what you look at, look up, and buy across the internet. Sometimes this information is even linked to offline purchases and activities you engage in.

Until I started writing this book, I had no idea that Comcast, my internet service provider, installed more than a hundred tracking tools.[28] Dictionary.com (one of my favorite websites, which I use more often than Facebook) installed 234 tracking tools on a user's computer without permission, only 11 from Dictionary.com itself and 223 from companies that track internet users.[29] The vast majority of these tools, according to a report by *The Wall Street Journal*, did not allow users to decline tracking. Among the 50 top sites assessed in this study by *The Wall Street Journal*, Dictionary.com "ranked highest in exposing users to potentially aggressive surveillance."

Increasingly ingenious and troubling technologies are used to learn ever more about you. Two apps on the iPhone and Android devices—Color and Shopkick—activate your phone's microphone and camera to collect background sound and light patterns from your location, be it a bar, your office, or your home. Using the same type of program that allows your iPhone to name a song based on just a few notes, Color makes assessments about your location to alert you if other people in your social network are nearby and Shopkick assesses if the store you've entered has a bargain to offer you. Silicon Valley blogger Mike Elgan points out the wealth of information marketers can collect about you through these phone apps: "Your gender, and the gender of people you talk to; your approximate age, and the ages of the people you talk to; what time you go to bed, and what time you wake up; what you watch on TV and listen to on the radio; how much of your time you spend alone, and how much with others; whether you live in a big city or a small town; what form of transportation you use to get to work."[30]

Browser cookies can be used by data aggregators to collect user IDs, user-selected preferences, demographics, purchasing histories, creditworthiness, log-in names, Social Security numbers, credit card numbers, phone numbers, and addresses.[31] How do they work? When a user types the web address (known as a URL) of a social network or website into a browser or clicks on a link for a website—as I did when I ordered books on Amazon—the browser contacts the website's server and requests the page.[32] The website's server then sends the requested web page to the browser. The website server treats each request as if it were the first request it had received from the user—the website server has no memory.[33] But when the website server places a small line of text—known as a cookie—on your computer, it can keep track of your subsequent visits to its web pages and actions you took on its website (for example, the title of the books you ordered on Amazon, as well as those you looked at but didn't order).[34] That information can be used to create personalized ads to sell you additional products in the future (such as other books of the genre you ordered).

A cookie can also be placed on a user's hard drive by a third-party advertiser. By 2001, the data aggregator DoubleClick had convinced 11,000 websites (including 1,500 of the most highly trafficked websites, such as AltaVista, *U.S. News and World Report* online, *The Wall Street Journal*, theglobe.com, NBC, *Reader's Digest,* and Bloomberg) to allow it to place cookies on their users' computers.[35] DoubleClick could then collect and aggregate data about what the user did on any of those 11,000 websites. DoubleClick used the information collected for behavioral advertising so their clients could decide which banners would be displayed when a particular person visited the Web. Here is an example of a DoubleClick cookie: id 80000008xxxxxxb doubleclick.net/ 0 1468938752 31583413 158986260829410552.*[36] And this is an example of a cookie from Hotmail in the Internet Explorer browser: HMP1|1|hotmail.msn.com/|0|1715191808|32107852|3 511491552|29421613|*|.[37]

A web beacon (also called a web bug, action tag, pixel tag, or clear GIF) can be used as an alternative means to compile information about an internet user. A web beacon is a small graphic image that is usually transparent (and therefore invisible to the user) and no larger than one pixel by one pixel that is placed on a website or in an email.[38] When an internet user visits a web page or opens an email containing a web beacon, the code of the web page or email instructs the computer to contact another server to download the web beacon.[39] This server is operated either by the owner of the website or by a third party that has permission to place the web beacon on the owner's website.[40] When the computer contacts the server to retrieve the small graphic image, the server generates a file about characteristics of the user, such as the internet protocol address (the unique address of the requesting computer), the web address of the page the person is viewing, the time the web beacon was loaded, and the type of browser that retrieved the web beacon.[41] This is an example of a DoubleClick web beacon hidden in the HTML code of Quicken: <img src="http://ad.doubleclick.net/ad/pixel.quicken/ NEW" width=1height=1 border=0>.[42]

Frequently web beacons and cookies are used in tandem. A web beacon can be used to deliver a browser cookie to the user's computer.[43] Through this method, a web server can recognize a browser across a number of domains and sites, permitting data aggregators to capture a user's web activity.[44]

Web beacons are everywhere. In a 2009 University of California at Berkeley study, each of the 50 most visited websites contained at least one web beacon, while most sites had several and some sites had as many as a hundred.[45] And some tracking companies have a wide scope of coverage. Google and its subsidiaries, for example, had web beacons on 92 of the top 100 websites.

Data aggregators also gather information via Flash cookies, which have been described as a "normal browser cookie on steroids."[46] Adobe Flash Player is software for viewing certain videos, animations, web apps, games, text, and images in internet browsers.[47] To support this, the Adobe Flash Player has its own storage system. Websites containing Flash applications can store information on an individual's hard drive. The data storage file, known as a Flash cookie, is used by websites to keep track of the user's preferences, such as the volume setting for a particular Flash application. But, just like browser cookies, Flash cookies have been co-opted by advertising networks and data aggregators to collect and store information about the browsing habits of internet users. And Flash cookies offer advertisers and data aggregators advantages over browser cookies, since they can store up to 100 kilobytes of information, while browser cookies can store only 4 kilobytes.[48] Flash cookies are also harder to get rid of than normal cookies. Erasing browser cookies, clearing browser history, erasing the cache, deleting private data within the browser, or changing the browser to "private browsing," which can remove or disable browser cookies, sometimes does not affect the Flash cookies.[49] Plus, deleted browser cookies can be "raised from the dead" by Flash cookies, creating "zombie" cookies.[50] A website server will place both browser and Flash cookies on a user's computer, and the Flash cookie will store the browser cookie's unique cookie ID. When the Flash cookie is activated, it will check for the existence of the browser cookie, and if the browser cookie does not exist because the user deleted it, the Flash cookie creates and installs another one.[51]

The most aggressive and problematic technology used for data aggregation and behavioral marketing is deep packet inspection. This technology allows internet service providers (ISPs) or third parties to collect and analyze the internet transmissions of an ISP's users.[52] Transmissions sent on the internet are broken into digital packets, each of which contains only a portion of the original transmission but all of which contain the internet protocol addresses of the sender and the recipient and an indication of that packet's place in the complete transmission. These packets are transmitted from router to router across the internet to their destination. Since some routers might be busy at a particular moment, sometimes packets are sent along different routes to the same destination.

As one judge explained it, "If a computer in New York sent a document to one in Boston, some packets might travel through routers and cables directly up the east coast while other packets might be sent by way of Seattle or Denver, due to momentary congestion on the east coast routes."[53]

Deep packet inspection by the internet providers themselves has a number of legitimate uses: detecting network attacks, managing network congestion, and charging different prices for different internet services.[54] But some behavioral marketing firms contract with ISPs to spy on—and copy—what users are sending.[55] The data aggregators place a tap on the ISP's equipment and collect and inspect all the packets of information sent out by users. This is a massive amount of data, including every email you send, every website you browse, voice-over-internet-protocol (VOIP) phone calls (such as through Skype), peer-to-peer file transfers, and online gaming. In her statement to the House Subcommittee on Telecommunications and the Internet in July 2008, Alissa Cooper, chief computer scientist for the Center for Democracy & Technology,[56] analogized deep packet inspection to a post office opening and reading a letter before it is sent.[57]

The data aggregator receives data packets from a person's transmissions and analyzes the content of the packets to create a profile of the person's online behaviors and interests. The aggregator can then sell the information and analyses to others, including advertisers who create targeted ads based on people's behavioral profiles.

When I did my Sunday morning business on social networks and websites, I had no intention to let others peek at—let alone sell—the information that could be gleaned about me from what I wrote, bought, sent, or viewed. "In part because the Internet was developed around the end-to-end principle, consumers have come to expect that their Internet communications pass through the network without being snooped on the way," says Cooper. "Deep packet inspection dramatically alters this landscape by providing an ISP or its partners with the ability to inspect consumer communications en route. Thus, deploying a DPI system likely defies the expectations consumers have built up over time."[58]

Even the games you play and the apps you use on Facebook can collect and transmit personal information about you. In 2007, Facebook launched a platform that let software developers build applications that run on the site. By 2011, there were more than 550,000 apps, and those apps have become an industry, with social games, the biggest category of apps, having a projected revenue of $1.2 billion annually.[59] Facebook reported in 2010 that 70% of its users run at least one app each month.[60]

A 2010 investigation by *The Wall Street Journal* found that many of the most popular applications on Facebook were transmitting identifying information about users and their friends to advertisers and internet tracking companies, which is a violation of Facebook's privacy policy.[61] *The Wall Street Journal* analyzed the ten most popular Facebook apps, including Zynga's FarmVille, with 59 million users, and Zynga's Mafia Wars, with 21.9 million users, and found that they were transmitting Facebook user IDs to data aggregators. When a data aggregator has a Facebook ID, it can access any public information on a person's Facebook page (which could include the person's name, age, residence, occupation, and photos). The Zynga applications were sharing Facebook users' IDs with the internet tracking company Rapleaf, which then added the information to its own database of internet users for enhanced behavioral advertising.[62]

Rather than focusing on an individual's interaction with a website, some data aggregators use a method known as "scraping" to extract all the data that anyone has posted on a particular website, analyze it, and sell it. Web scrapers copy information from websites through specially coded software.[63] These software programs are also referred to as web robots, crawlers, spiders, or screen-scrapers. Scrapers are designed to search through the HTML code that makes up a website and extract desired information. If a certain website includes a discussion by new moms (or by people considering buying cars), the data scraper can sell that information and the people's email addresses or IP addresses to advertisers who want to target ads to those types of consumers.

Web scrapers "are capable of making thousands of database searches per minute, far exceeding what a human user of a website could accomplish," says attorney Sean O'Reilly, who previously worked in the software industry. "Web vendors have a difficult time detecting a difference between consumers accessing this information for their own benefit, and aggregators accessing the information to return to their own databases."[64]

Search engines such as Google, Yahoo!, and Bing also collect, store, and analyze information about individual users through their search queries. Search engines maintain "server logs," which, according to Google's Privacy Policy, include your "web request, Internet Protocol address, browser type, browser language, the date and time of your request and one or more cookies that may uniquely identify your browser."[65] Microsoft's search engine, Bing, adds that it also "will attempt to derive your approximate location based on your IP address."[66] Search engines use this information to optimize their search algorithms and to record an individual's preferences.[67] Though Google uses these logs for fraud prevention and to improve search results, it also analyzes the logs to generate more revenue through targeted advertising.[68] Yahoo! also uses this information to personalize advertising and page content. Yahoo! acknowledges that it also allows other companies to display ads on its pages and those ads may "set and access cookies on your computer" that are not subject to Yahoo!'s privacy policy.[69]

In 2006, AOL made public 20 million queries entered into its search engine from 658,000 users on its website research.aol.com.[70] AOL's release contained all of those users' searches over a three-month period and detailed whether they clicked on a result, what the result was, and where it was in the list of results.[71] An AOL researcher, Abdur Chowdhury, explained the release of the queries as an effort to facilitate "closer collaboration between AOL and anyone with a desire to work on interesting problems."[72] But the project ended up breaching people's privacy. In some instances, people could be identified through the types of searches they undertook.

A quick look at some of the leaked AOL search logs makes it easy to imagine how damaging a search log can be when linked to a party in a criminal, civil, or divorce case.

User 11574916:
cocaine in urine
asian mail order brides

states reciprocity with florida
florida dui laws
extradtion from new york to florida
mail order brides from largos
will one be extradited for a dui
cooking jobs in french quarter new orleans
will i be extradited from ny to fl on a dui charge

User 336865:
sexy pregnant ladies naked
nudist
sexy feet
child rape stories
tamagotchi town.com
preteen sex stories
illegal child porn
incest stories
10 year old nude pics
preteen nude models
illegel anime porn
yu-gi-oh

User 59920:
cats skinned in fort lupton co
cats killed in fort lupton co
jonbenets autopsy photos
crime scene photos of the crawl space and duffle bag in ramseys house
sexy bathing suits
what a neck looks like after its been strangled
pictures what a neck looks like after it was strangled
pictures of murder victims that have been strangled
pictures of murder by strangulation
knitting stitches
what jonbenet would look like today
new jersey park police
jonbenet in her casket
ransom note in the movie obsession what did it read
movie ransom notes
scouting knots
manila rope and its uses
brown paper bags cops use for evidence
rope to use to hog tie someone
body transport boulder colorado

User 1515830:
  chai tea calories
  calories in bananas
  aftermath of incest
  how to tell your family you're a victim of incest
  pottery barn
  curtains
  surgical help for depression
  oakland raiders comforter set
  can you adopt after a suicide attempt
  who is not allowed to adopt
  i hate men
  medication to enhance female desire
  jobs in denver colorado
  teaching positions in denver colorado
  how long will the swelling last after my tummy tuck
  divorce laws in ohio
  free remote keyloggers
  baked macaroni and cheese with sour cream
  how to deal with anger
  teaching jobs with the denver school system
  marriage counseling tips
  anti psychotic drugs[73]

Your web searches provide data on which you can be judged, erroneously or not. If you've looked up the side effects of antidepressants, that information might be used against you by an employer or a college admissions officer. Your search for a divorce lawyer, advice about green cards, or information about sexually transmitted diseases might also be used in ways that harm you.

Your second self on the Web is likely a distortion of your offline self. The person whose leaked AOL searches related to extradition might have been writing a mystery, rather than covering up a crime. The woman who was seeking information on AOL about incest might have been trying to help a friend, rather than dealing with her own troubled past.

When AOL released the supposedly anonymous queries, it was easy for reporters from *The New York Times* to identify Thelma Arnold as searcher 4417749 due to her searches for other Arnolds and her searches about Lilburn, Georgia.[74] After discussing her queries for 60-year-old single men, queries about her three dogs, and queries researching her friends' ailments, Thelma said, "My goodness, it's my whole personal life. I had no idea somebody was looking over my shoulder."[75]

But "in user search query logs, what you see is not always what you get," notes Omer Tene, a professor at a law school in Rishon Le Zion, Israel. Anyone who had access to Thelma Arnold's logs saw searches for "hand tremors," "nicotine effects on the body," "dry mouth," "bipolar," and "single dances in Atlanta." However, those were searches Thelma conducted for others and do not paint an accurate picture of her life or health.[76]

The attributes of your digital doppelgänger may have more influence on what opportunities you receive than any of your offline characteristics. Rather than expanding opportunities for you, the targeted ads that you see may actually deny you certain benefits. You might be shown a credit card with a lower credit limit, not because of your credit history but because of your race, sex, zip code, or the types of websites you visit. As a consequence of weblining, the information collected by data aggregators is often sold to the public at large (through websites such as Spokeo) and might later hamper your efforts to get a job, qualify for a loan, adopt a child, or fight for your rights in a criminal trial.

As behavioral advertisers increasingly dictate a person's online and offline experiences, stereotyped characterizations may become self-fulfilling. Rather than reflecting reality, behavioral analysis may inevitably define it. When young people from "poor" zip codes are bombarded with advertisements for trade schools, they may be more likely than their peers to forgo college. And when women are routinely shown articles about cooking and celebrities, rather than stock market trends, they will likely disclaim any financial savvy in the future. Behavioral advertisers are drawing new redlines, refusing to grant people the tools necessary to escape the roles that society expects they play. Our digital doppelgängers are directing our futures and the future of society.

Some social network users feel that access to their personal data is a small price to pay for the ability to use Facebook and other websites without charge. Other people, though, do not want to face discrimination based on their second self and want to use technological and legal measures to assert control over their own data. Still others feel that information about them should be considered to be their property and if anyone is going to make money by selling that data, it should be them. Yet without a Social Network Constitution, individual choices go unrecognized as the networks and data aggregators call the shots.

A shift away from third parties piecing together and shaping our second selves is imperative to avoid predetermined fates. We must be given an opportunity to create our own alter egos and paint a unique, personal self-portrait. A Social Network Constitution will help us do just that. It can restore power to the people and open up the possibility for individuals to once again explore their identities and determine their fates.

WEB TRACKING CHART

	Deep-Packet Inspection	Scraping	Flash Cookies	Browser Cookies	Search Engines	Remotely Installed Keylogger
Sensitive information submitted through web forms (e.g., credit card info, Social Security number, passwords)	Yes	No	Yes	Yes	No	Yes
Skype calls (to whom, for how long)	Yes°	No	No	No	No	No
Websites visited in a session	Yes	No	Yes	Yes	No	No
How long you've stayed on a page	Yes	No	Yes	Yes	No	No
Your IP address when you access a page	Yes	No	Yes	Yes	No	No
Facebook postings (public)	Yes	Yes	No	No	Yes	Yes
Facebook postings (private)	Yes	Yes†	No	No	No	Yes
Contents of private forum posts (e.g., posts on PatientsLikeMe)	Yes	Yes‡	No	No	No	Yes
Content of sent emails	Yes	No	No	No	No	Yes
Content of attachments in sent emails	Yes	No	No	No	No	No
Ticket purchase info on airline site	Yes	No	Yes	Yes	No	No
Info submitted to government site to obtain license	Yes	No	Yes	Yes	No	Yes
Browsing and purchasing activity on Amazon.com	Yes	No	Yes	Yes	No	No
Clicks through to a website via email link	Yes	No	Yes	Yes	No	No
Search query about a town	Yes	No	Yes	Yes	Yes	Yes

	Deep-Packet Inspection	Scraping	Flash Cookies	Browser Cookies	Search Engines	Remotely Installed Keylogger
Download of a newspaper article (plus possibly the title from the URL)	Yes	No	Yes	Yes	No	No
Doctor email changing prescription electronically	Yes	No	No	No	No	No
Facebook activity (that doesn't require typing)	Yes	No	Yes	Yes	No	No
Untagging personal photo	Yes	No§	No	No	No	No

*However, even if the interception of voice traffic is technically possible, the compressed data will likely be encrypted.

†Yes, though collectable data is limited to accounts with which the scraper's account is friends.

‡Having an account is necessary to access private forums.

§Prior to untagging, scrapers with Facebook accounts that can view the image will also be able to collect the data linking you to the image.

Thanks to Cynthia Sun for the preparation of this chart.

3

Second Self

Concerned by her four-year-old son's worsening strep-like symptoms, Deborah Copaken Kogan posted photos of the child and updates on Facebook: "Eyes swollen shut. Fever rising. Penicillin not working. Might be scarlet fever. Or roseola. Or . . . ????" and later, "Swelling worse . . . especially eyes and chin. Fever still crazy high." The posts elicited responses from three Facebook friends (including an actress whose son had the condition and a pediatric cardiologist) all urging her to take the child into the hospital immediately since they suspected that the child had Kawasaki disease, a rare auto-immune disorder that can destroy coronary arteries if untreated. The boy was in fact diagnosed with Kawasaki, leading the mother to credit Facebook with saving her son's life.[1]

When Bobbette Miller needed a kidney, Donnie Wahlberg of New Kids on the Block was able to elicit numerous offers for kidney donations by simply tweeting a post about her plight to his 183,000 Twitter followers. Donnie's actions led to six matches for Bobbette, who underwent successful transplant surgery in June 2011.[2]

Social networks expand the circles of care and of information. Patients who live in remote areas or suffer from rare diseases can come together in a private space to augment their medical care by hearing from others with similar symptoms or concerns. Through the social network PatientsLikeMe, for example, people share their health information in hopes of learning from others and better managing their own treatment.[3] They discuss potentially stigmatizing conditions, including schizoaffective disorder, brain damage in infancy, eating disorders, borderline personality, Parkinson's disease, panic disorder, hemorrhoids, drug addiction, and social anxiety disorder.[4] PatientsLikeMe members feel secure in disclosing such intimate information because they can use a pseudonym instead of their real name and they can choose to allow their posts to be read only by PatientsLikeMe members, instead of the public at large.[5]

Thirty-three-year-old Bilal Ahmed joined PatientsLikeMe.com to connect to other sufferers of depression.[6] Bilal posted on the password-protected "mood" discussion forum and listed the medicines he used. In May 2010, he discovered that the data aggregator Nielsen (which has pharmaceutical company clients) had posed as a new member of the site and used data-scraping software to copy messages from the website's private forums.[7] Bilal removed all of his posts from PatientsLikeMe. "I felt totally violated," he told *The Wall Street Journal*. "It was very disturbing to know that your information is being sold."[8]

Even though Bilal had used a pseudonym on the site, he was identifiable through a link to his personal blog. Thus, the Nielsen Company knew not only his most intimate medical information but his actual identity. Other profiles could also potentially be traced to specific individuals. One user, 1955chevy, provided his age, the small town in the South where he lives, and the Home Depot department he works in. He was diagnosed with multiple sclerosis in June of 1998 and lists his secondary disorders as polymyalgia rheumatica, hypothyroidism, and shingles. He includes a list of his prescription drugs, supplements, general symptoms, condition-specific symptoms, relapses, and a graph of his weight. 1955chevy also uses a PatientsLikeMe function that allows him to comment on his social, mental, and physical quality of life.[9]

Another user, IAmDepressed, is a 15-year-old male who provided the name of the small town in the Northeast where he lives. IAmDepressed's profile describes his two-year struggle with depression, including his suicidal thoughts, treatment history, and medicines. He also notes that telling his mother about his depression "didn't really open her eyes enough. It was a little shrugged off."[10] His profile features a "Mood Map," which includes charts of his mood function, sleep, appetite, anxiety, and emotional control.

Before the incident, the people on PatientsLikeMe shared information about depression, spoke of thoughts of suicide, and provided advice about drug protocols that helped them. Once the break-in occurred, some people started deleting their information from the site and stopped posting, which undercut the benefits of social networks to people with health conditions.

This incident shows how far the norms of privacy on social networks have strayed from the norms in the rest of our lives. Off the Web, federal and state laws recognize the privacy of an individual's medical information since, if it falls into the wrong hands, it might lead to stigma and discrimination. But those laws, including the privacy regulations adopted under the Health Insurance Portability and Accountability Act,[11] apply only to specifically defined "covered entities," which include health care providers and health plans but not social networks.

Data aggregators know a huge amount about you from your interactions on Facebook and websites, but they don't have a context for it. A person on PatientsLikeMe might post about her anxiety and then do a search for life insurance. A data aggregator, assuming that the life insurance was for her, might use these two bits of information to put the person into the category of someone likely to commit suicide. Subsequently, when the woman goes on a bank website, she might only be able to obtain a credit card with a low limit and might not be offered any loans since she might be categorized as likely to kill herself and thus a bad credit risk. In reality, though, the woman might have been anxious because she's in the final stages of setting up a new business and she could have been searching for insurance for the employees she plans to hire.

The massive acquisition of information by data aggregators, their analyses of it, and their sale of it not only invades your privacy but also denies your individuality and can harm you emotionally and financially. Your private data is being used not just to sell you products but to deny you certain opportunities.

Data aggregators slot you into a category and make assumptions about you based on that category. The New York–based company Demdex builds "behavioral data banks" that contain information collected from people's browsing history and retail purchases.[12] Demdex then puts people into categories such as "midlife-crisis male" or "young mother" to determine what type of ads should appear when a person visits a retailer's website.[13]

The data aggregator Acxiom is in the business of collecting, analyzing, and selling personal information.[14] Acxiom has data on half a billion people from around the world, including 96% of Americans. The company has an average of 1,500 pieces of data on each person, notes Eli Pariser in *The Filter Bubble: What the Internet Is Hiding from You*, "everything from their credit scores to whether they've bought medication for incontinence."[15]

Acxiom assigns you a 13-digit code and puts you into one of 70 "clusters" based on your behavior and demographics, including your household characteristics such as age, income, and net worth, whether there are children in your home, and whether the area you live is urban, suburban, or rural.[16] People in Cluster 38, notes Acxiom's website, are most likely to be African American or Hispanic, working parents of teenage kids, and lower middle class and shop at discount stores. Someone in Cluster 48 is likely to be Caucasian, high-school educated, rural, family oriented, and interested in hunting, fishing, and watching NASCAR. Within Cluster 26, is a group of singles averaging 37 years old in the upper-middle income ranges. They "attend professional sporting events," "participate in various indoor and outdoor sports such as weight lifting, going to the fitness club, mountain biking and tennis," and "with their spending on MP3 players, it looks as though the tunes always go along for the ride."[17]

Acxiom's website allows you to infer what cluster you are in.[18] When a blond Midwestern law student filled in the required questions—her age, marital status, the fact she was a renter, her combined household income, zip code, and net worth, Acxiom stuck her into Cluster 61, which has a high concentration of Asians, Hispanics, and African Americans. Acxiom guessed right that she had a "below average income" but wrongly suggested she enjoyed going to movies. Because it erroneously assumed she must be a member of a minority group, Acxiom put her into a cluster in which her "strong interest in foreign travel is most likely driven by visits to family abroad."

Data aggregation and behavioral advertising run contrary to some fundamental social values. Under the U.S. Constitution and civil rights laws, institutions are supposed to make decisions on people based on their individual characteristics, not because of aggregated data that assumes that people who are in a certain demographic are more likely to do X or Y. Nor can they profile people based merely on race. Realtors can't refuse to show a person a house just because he or she is a member of a racial group that is less likely to be able to afford that house. A person must be judged on his or her own merits. Cops need to have suspicion about a particular individual before they search him for a concealed weapon. Even if, in a particular group or a particular neighborhood, there's a very high chance that people will be carrying an illegal weapon, that aggregate data doesn't give cops the authority to search every person in that group or neighborhood without individualized suspicion.

The right to be treated as an individual, rather than a member of a group, is fundamental in a democratic nation. But that right is turned on its head with social networks' data aggregation and behavioral advertising. Real harms can occur. The surreptitious collection and use of online information can lead to psychological risks, financial risks, discrimination risks, and social risks. Yet digital snooping laws provide little protection. The few laws that exist were enacted to deal with earlier technologies, such as the telephone, or require proof of a high level of financial loss before a person can sue. Even when laws seem to be on point, they've often been stripped bare by judges who privilege the development of technologies over individual rights or who create exceptions to facilitate law enforcement, unintentionally providing loopholes for data aggregators, too.

The information collected on your second self can entrench discrimination and deny you benefits. The collection of potential customers' personal information for behavioral advertising gives companies greater ability to deny opportunities to people based on that personal information or their location.[19] When people logged into Wells Fargo's website to learn about houses for sale, the site collected zip code information and used it to direct potential buyers towards neighborhoods that were of a similar racial makeup.[20]

The company [x+1] uses aggregated online data to make snap assessments of individual visitors to websites. Businesses use [x+1] to determine which advertisements and what content appears when someone visits their websites. For example, Capital One uses [x+1] information to instantly determine which credit cards to show to first-time visitors to its website. Clients pay [x+1] an estimated $30,000 to $200,000 a month for this type of service.[21]

The Wall Street Journal had individual testers visit Capital One's site, and [x+1] reported what identification resulted. The [x+1] software correctly determined that: Carrie Isaac is "a young Colorado Springs parent who lives on about $50,000 a year, shops at Wal-Mart and rents kids' videos" (her Capital One page displayed less generous credit cards); Paul Boulifard is "a Nashville architect, is childless, likes to travel and buys used cars" (his Capital One page displayed a travel rewards credit card); and Thomas Burney is "a Colorado building contractor, is a skier with a college degree and looks like he has good credit" (his Capital One page displayed a very generous credit card with an initial 0% interest rate and no annual fee).[22]

For the law school course I taught on social networks, I asked students to do a simple in-class experiment. Everyone went to a neutral-seeming website, www .tvguide.com, and compared the ads that popped up on the home page. Even though the students were roughly the same age and pursuing the same profession, there was a great variation in the ads. Men received ads for expensive cars and higher-limit credit cards. Women received ads for rental cars or ads about starting a family. Gender and race discrimination are blossoming in the arena of behavioral advertising.

Even something as basic as customer service can be limited by your digital doppelgänger. When you call to ask a company a favor or make a complaint, the action you get in response is often influenced by how a data aggregator has characterized you. At some banks, whether your bounced check fees are waived will depend on that categorization.[23] Recently, I called Comcast to report a service problem. An automated response asked me to key in my phone number, telling me I'd get a call back in 20 minutes. My second self must rank pretty low in Comcast's eyes, since nearly a week has passed and Comcast still hasn't called back. If I were in line in a bank or a store, I'd notice if people were called out of order or if all the men received better service than any of the women. But on the internet, weblining can occur without your even knowing it.

When Marcy Peek, an African-American lawyer, first heard of weblining a decade ago, she wrote, "The ultimate problem is not just that online profilers intrude into our privacy zones and survey our personal space via 'inherently unfair and deceptive' methods, nor is it just that, as at least one commentator has argued, with the taking of our personal information they take from us a feeling of dignity and humanity. Rather, in the new paradigm, corporations and those holding the power of information distribution and access have the technological capability to determine who has access to information, opportunities, financial offers, and certain economic advantages."[24] Peek had a unique solution—she advised people to "pass"—to construct their identity as if they were a member of the favored group.

Hers was a fascinating idea and might have worked when she proposed it in 2003. But since then passing has become virtually impossible. With the birth of social networks, one's name is disclosed and one's race and gender are obvious through profile pictures and the aggressive actions of data aggregators. Now the financial harms of weblining are endemic, indicating a need for social solutions rather than individual ones.

Beyond potentially losing out on benefits, people can be emotionally harmed by the way behavioral advertising plays to their worst fears or facilitates self-destructive behavior. Seventeen-year-old Cate Reid was worried about her weight. Wishing to lose 15 pounds, she searched for weight loss information online. After that, every time she went online, whether she was seeking out information about weight loss or not, she saw weight loss ads. These ads kept bringing her weight to the front of her mind. "I'm self-conscious about my weight. I try not to think about it. . . . Then [the ads] make me start thinking about it," Cate told *The Wall Street Journal*.[25]

Julia Preston, a 32-year-old education software designer living in Austin, Texas, had a similar experience that she found to be "unnerving."[26] Julia had researched uterine disorders online. After her research she started noticing fertility ads on websites she visited, since some women with uterine disorders have problems conceiving. Even though she didn't have a uterine disorder, the ads kept appearing.

Whitney Chianese, a 28-year-old from Rye, New York, was exchanging emails with her mother in Atlanta. Whitney's grandmother had just recently died, and Whitney and her mother were discussing her death. While Whitney was online, pop-up advertisements began to appear for health care products. Whitney said, "It was like Big Brother. It became too much."[27]

Cate, Julia, and Whitney's stories show the emotional consequences of behavioral advertising. The Yahoo! ad network had collected information about Cate's online behavior and formed a profile that categorized her as a recent high school graduate and a "13- to 18-year-old female interested in weight loss."[28] Julia was targeted by Healthline Networks, an advertising network that "scans the page a user is viewing and targets ads related to what it sees there. So, for example, a person looking up depression-related words could see Healthline ads for depression treatments on that page—and on subsequent pages viewed on other sites."[29] Google uses its AdWords system to generate ads based on the keywords used in Gmail, in searches, in saved Google Chats, or on discussion pages,[30] and used its scan of Whitney's emails about her deceased grandmother to target ads.

In a particularly macabre twist on behavioral advertising, pop-up ads sometimes appear on social network websites devoted to suicide. If a person says she is going to overdose on a certain type of pill, an ad might appear saying "Call this 800 number immediately for a discount on those pills." On September 25, 2010, Google admitted that its automated system for advertisements had run advertisements for poisons and chemicals on the Google Group page alt.suicide.methods,[31] where users discussed how to do themselves in.[32] Advertisers select keywords for their advertisements in order to narrowly target consumers.[33] So when Joanne Lee and Stephen Lumb used the forum to arrange a suicide pact by means of a gas-filled car, the forum page featured the following advertisements: "Sulphuric Acid. Call us free on 0800 090 ****" and "Hydrogen Sulphide. Find medical & lab equipment. Feed your passion on eBay.co.uk!"[34]

The fact that behavioral advertising occurs under the radar not only creates the possibility of individual harms, it also fosters societal harms. In fact, behavioral advertising may have precipitated the subprime mortgage crisis. Mortgage lenders Countrywide Financial and Low Rate Source were two of the 10 biggest online advertisers in the United States in July 2007.[35] Google and Yahoo! were among the primary beneficiaries of the subprime mortgage market in 2007. They "cashed in big time from the mortgage boom," said financial blogger Faisal Laljee. "Direct lenders, conventional banks and lead aggregators like Lending Tree, Nextag, and

LowerMyBills.com have all paid top dollar to drive online traffic to their site."[36] During peak times Google was selling ad space next to its search engine for search queries including keywords such as "mortgage" and "refinance" for as much as $20 to $30 per click on the placed ad.[37]

Google, Yahoo!, and MSN, along with financial websites (such as Bankrate .com and Mortgage War) and behavioral advertising firms (such as 24/7 Real Media and Revenue Science) also sold information about internet users identified as likely prospects for a mortgage.[38] The internet leads were sold to mortgage companies, which then contacted people who could not get loans from their local lending institutions.[39] The customers targeted through these internet leads for subprime mortgages were predominantly low income, black, and Hispanic.[40] And they usually got a bum deal. A national study of subprime loans conducted by the Center for Responsible Lending found that people of color were 30% more likely to be charged higher interest rates than white borrowers with the similar credit ratings.[41] When borrowers defaulted on the subprime mortgages, the entire economy was affected.[42] Subprime lenders lost money and were unable to service new loans, inevitably going out of business.[43] Investment banks that had traded in securities backed by bad mortgages, such as Bear Sterns, Lehman Brothers, Goldman Sachs, Merrill Lynch, and Morgan Stanley, also suffered significant financial losses, which were reflected in a plummeting stock market.[44]

The data aggregation business keeps a low profile but has high earnings. Three years after DoubleClick went into business, it was wealthy enough to spend over $1 billion to acquire a direct marketing company that had a database of the names, addresses, telephone numbers, retail purchasing habits, and other personal information of approximately 90% of U.S. households.[45] In 2007, Google and Microsoft got into a bidding war for DoubleClick, with Google ultimately paying $3.1 billion to purchase the company.[46] Rather than Big Brother watching, it is Big Business that is monitoring, recording, analyzing, and even selling your private information.

So how can people take action against surveillance and stereotyping due to data aggregation? Some data aggregation companies offer to let users opt out of being tracked. But if you don't know that Spokeo or DoubleClick is collecting information about you, it is doubtful that you would think to track down their websites and figure out how to opt out. And for many websites, the opt-out option is illusory. Often websites cannot load or function properly if you don't accept their tracking technologies. So although web browsers can be set to not automatically accept cookies, you may not be able to access a website at all if you don't accept the cookies, web beacons, or other data collection mechanisms. Either the site will not load, or you will spend an inordinate amount of time responding to tracking requests in order to load the site. (See "A Law Student's Attempt to Opt Out.")

And sometimes the opt-out doesn't even work as promised. Facebook said it would let users opt out of a certain type of web beacon, but when a code-savvy journalist opted out, he discovered that Facebook was still collecting his data.[47] When people used the opt-out alternative on the site of the data aggregation firm Chitika, they were not told that the company was going to opt them out for a mere ten days and then reinstall all the cookies and beacons after that time.[48] I dutifully followed the annoying seven steps to remove my name and data from the Spokeo website. But when I searched Spokeo again a few months later, I discovered that Spokeo was listing information about me again. Aperture has a particularly perverse opt-out approach. A division of Datran Media, Aperture combines "household-level demographic data verified by multiple offline third-party sources" with behavioral data based on interests and "transactional-based behavioral data based on real time conversations."[49] In order to opt out, an internet user must visit Aperture's website and click the "Opt-Out" link located in the banner. The user then receives a notice stating that although the user has opted out, if she clears her cookies, she must re-visit the web page to opt out again.[50] So the very attempt to clear cookies backfires and reinstates tracking by Aperture!

A LAW STUDENT'S ATTEMPT TO OPT OUT

To manage my cookies, I first went to the Tools, Internet Options, Privacy Settings, Advanced tab on my Internet Explorer 8 Browser. I requested that it block all first-party and third-party cookies. I quickly found out that with that setting Gmail and Facebook would not allow me to get past the log-in. I then changed the settings to "prompt" me when a site wanted me to accept a cookie. This was similarly difficult. At Gmail I accepted seven cookies, seemingly all from Gmail or Google themselves, but even after accepting these cookies, my log in was not successful and a warning came up that Gmail requires cookies to be enabled. I then typed Facebook.com into my browser. I had to accept four cookies before the log-in screen came up. After typing in my username and password, I had to accept another eight cookies before I could get to the site itself. Based on the cookie names, they were all from Facebook itself. As I clicked around the Facebook site, I received requests to accept third-party cookies. A few of these prompts came up with almost every action I took. If you have set your privacy settings to "prompt," the prompt gives you the option of always allowing or always blocking cookies from a certain site. After choosing to always allow cookies from Google and Facebook, I was allowed to get into my Gmail and Facebook accounts.

Elizabeth Raki

Some people seek technological solutions, but so far there is no all-purpose software or hardware you can install to thwart every single data aggregator. Every time progress is made with a technology to rid your computer of one type of assault (such as browser cookies), another type is launched to get around it (such as Flash cookies, which are more difficult to evade and can resurrect deleted browser cookies).

Some of these surveillance technologies piggyback on the legitimate products of others (such as the Flash cookies installed by Adobe to set volume levels in videos). As a result, the companies who sell the legitimate product have an incentive to develop technological fixes to protect their customers. Adobe began working with several companies, including Google and Mozilla, to enable users to delete Flash cookies from their browser history.[51] In 2011, Microsoft announced that users of the Internet Explorer 8 or 9 browser and the latest version of Adobe Flash (version 10.3) could delete Flash cookies from their browser history by using the "delete browser history" feature in the Internet Explorer browser.[52] Google Chrome now provides a way to delete Flash cookies. Mozilla's Firefox browser developed an add-on that can be loaded to help delete Flash cookies.[53]

But far more money and technological savvy has been invested in the development of data collection and aggregation mechanisms than has been invested into techniques to do away with them. And the power and scope of the tracking technologies just keep expanding. In 2009, PeekYou (which operates the website PeekYou.com, where you can find private information about people) filed for a patent application on a technology that would be able to collect information about people from anywhere on the Web—including social network pages—using scrapers. The purpose is to create a "comprehensive directory of all Internet users that tracks every individual's online presences."[54]

The patent says that PeekYou would collect information about each person from the following websites, among others—YouTube, Meetup, eBay, InfoSpace, Switchboard, PlentyOfFish, TagWorld, Faceparty, RateMyTeachers.com, RateMy Professors.com, Forbes, WAYN, Twitter, LiveJournal, Xanga, Yahoo!, Myspace, Friendster, Flickr, Bebo, LinkedIn, and hi5. The patent application provides an example: an image of John Doe taken at a local restaurant will be scanned for data, such as geographic location (latitude, longitude, and altitude), exposure, date and time, resolution, and camera make and model. The data scraper would then scan the database for an individual's profile that matches the geographic information with a high degree of certainty. For example, if the digital photo was taken close to the user's home, there is a high likelihood that the image is associated with the user.

The PeekYou patent application describes many possible uses for its personal information aggregator. The addresses of incoming emails could be automatically scanned and the information contained in the database about the sender displayed for the email recipient. This distributed personal information aggregator could be combined with a geographic location device, such as a GPS-enabled cell phone, so that a person walking down the street could use his smartphone to view the profile information of strangers walking nearby.

In dealing with data aggregators, some people feel the best defense is a good offense. Ordinary individuals are teaming up with companies that offer to pay people for the types of data that the aggregators take without consent. A London company, Allow Ltd., lets people choose what information about them goes to behavioral advertisers and gives the individual 70% of the money earned by selling the information. "I wouldn't give my car to a stranger" for free, a London real estate developer, Giles Sequeira, told *The Wall Street Journal*. "So why do I do that with my personal data?"[55] By releasing certain information about himself through Allow Ltd., he gets paid. But he's also creating an incentive for the company to make sure that information is not collected without his consent by other data aggregators, since that would diminish Allow's resources as well as his.

Companies have also started offering to clean up negative information about people that data aggregators or others have made available on the Web. Although Reputation.com (formerly called ReputationDefender)[56] is the most prominent of the online reputation companies, there are many others in the marketplace, including Internet Reputation Management, Reputation Hawk, Netsmartz, and ReputationDR.[57]

The World Economic Forum selected Reputation.com as a Technology Pioneer for 2011, listing it among companies whose "technologies and business models will have a durable and valuable effect in several industries and societies as a whole. We look forward to their unique contributions to the mission of the Forum: improving the state of the world."[58]

But what can the reputation defender community do? Most commonly, people use it either to attempt to get problematic information or photos about them removed from the Web or to overwhelm negative information with positive information. A first-time visitor to Reputation.com is invited to search for his name to see what information about him exists on the Web. The search returns aggregated results that are similar to those found on people-search engine sites like Spokeo.com and PeekYou.com, such as an individual's location, photos, and work information.

Reputation.com charges $14.95 per month for a detailed monthly report that indicates what is being said about you online and $29.95 for each item you want removed from the Web,[59] with no guarantee of success.[60] When a member requests that an item be removed from a website, Reputation.com sends repeated letters to the website operators requesting the item's removal, escalating the language as needed.[61] In one case, a family paid $3,000 to ReputationDefender to try to scrub the photos of their daughter's corpse from the internet.[62] ReputationDefender successfully had the photos removed from around 300 sites, but the photos are still available on the Web on at least 100 sites.[63] "There is no silver bullet," says Michael Fertik, Reputation.com's founder. "If there was, I'd be a billionaire already."[64]

Reputation.com also made some headway in combating distressing and libelous comments about two female law students when, for no reason, they were ganged up on by posters from around the country. But the company has come under criticism for entering what it says are "exclusive removal agreements with some of the largest people databases on the Web."[65] In fact, Spokeo and PeekYou both have links to Reputation.com in the privacy sections of their websites and apparently are compensated for individuals they direct to Reputation.com's services. David Lazarus, a columnist for the *Los Angeles Times*, is critical of these business partnerships, since the people-search and privacy-protection sites should be at odds with each other, not partnering to help grow each other's profits. "Because Spokeo gets a cut of the action any time a user signs up through its site with ReputationDefender, it finds itself in the interesting position of profiting from a solution to the problem it helped cause," wrote Lazarus.[66]

The best way to counter the harms from data aggregation would be to prevent them from occurring in the first place—through precedential lawsuits or new legislation. Many people have sued data aggregators for violating federal and state laws, but courts have created loopholes in the federal laws that privilege data aggregators over the people whose data they collect. State laws on privacy or computer hacking may offer a possible alternative, but we may need to turn to novel legal theories, such as by giving people a property right over their own data.

There are several federal laws related to computer hacking and the regulation of electronic communications that could be applied to the unauthorized collection of information about people from social networks and other parts of the Web. The Computer Fraud and Abuse Act, the Stored Communications Act, and the Wiretap Act could have been interpreted to protect people. Courts, however, generally have refused to use these laws to protect against the collection of personal information through cookies or the interception of transmissions from a person's computer.[67]

The Computer Fraud and Abuse Act[68] makes it illegal to intentionally access a protected computer without authorization and transmit information, cause damage, or obtain information. But a person can't sue under the law unless he or she has suffered at least $5,000 worth of damage. In my Sunday morning computer browsing, I can't prove that the information collected about my purchases from Amazon.com and Southwest Airlines, my medication information, and other data I sent, received, or viewed is worth $5,000 (even though the disclosure of the information might cause me to lose my job or another benefit worth more than $5,000). And the fact that the cookies and web beacons might cause my computer to run more slowly is not considered a harm to me worth $5,000.

The Stored Communications Act[69] might seem like a better bet for a lawsuit. It makes it illegal to knowingly and unlawfully access a facility providing storage of electronic communications and obtain, alter, or prevent another authorized user from accessing the communication. Surely, that must provide protection against data aggregators that knowingly put a cookie on my hard drive to obtain information about me. The Stored Communications Act contains an exception to liability when there is consent to the access of the stored communications. That provision, too, makes sense. I shouldn't be able to sue Amazon for using my credit card to bill me for a book if I consented to its access to my credit card information for that very transaction.

But courts have distorted the consent requirement beyond recognition. They have decided that if Amazon consents to a marketing company secretly putting a cookie on my computer, I can't sue that third party under the Stored Communications Act.[70] The courts have held that one party's consent is enough. But shouldn't it be the consent of the person whose personal information is being collected that is considered—not the consent of the entity profiting from that surreptitious activity?

Another federal law, the Wiretap Act, makes it illegal to intercept communications and intentionally disclose or use them.[71] That law seems to be the perfect avenue for thwarting the data aggregators who use deep packet inspection or use websites to plant cookies. But the Wiretap Act also contains a consent exception and, once again, courts have considered the consent of the website, which is being paid by the data interceptor, as sufficient. According to the courts, the consent of the person whose data is being intercepted is not required.

Think about the absurdity of one-party consent. If a rival hires someone to beat me up, should it matter that he has given his consent to the act if I haven't? In passing these laws, Congress recognized that it makes no sense to have one-party consent in situations in which the entity giving the consent is accessing the computer or the transmission to commit a tort or crime. If that weren't the case, a hacker copying my Social Security number from my computer would not be committing a crime if he himself had consented to invading my computer.

In the landmark 2001 case *In re DoubleClick*, a New York federal judge was asked to find a data aggregator in violation of the federal laws because, even though the website had consented, the collection was being done as part of a tort (invasion of privacy) and crime (trespass into the computer). In a decision that should be called Double Speak instead of *DoubleClick*, the judge said that the data aggregator's intent wasn't to commit a tort or a crime, but rather to make a lot of money so its activities were permissible. In the court's words, the company was "consciously and purposefully executing a highly publicized market-financed business model in pursuit of commercial gain."[72]

If someone broke into my house and put a videocam in my bedroom, would we really let him get away with it if he said, "I wasn't intending to invade your privacy, I just run a business where we sell sex tapes"? Or would we let someone who embezzled funds say, "I wasn't intending to drain everyone's pensions, I was just intending to gather as much money as possible"?

In an equally unpersuasive case in 2010, a New York state judge refused to apply a criminal statute regarding unauthorized use of a computer because the judge likened a person's electronic communications (in that case, an email) to a "postcard" and did not consider them to be private.[73]

"Moreover," said the judge, "emails are easily intercepted, since the technology of receiving an email message from the sender, requires travel through a network, firewall, and service provider before reaching its final destination, which may have its own network, service provider and firewall."[74] But my landline phone calls likewise go through various networks to reach a destination, yet laws prevent people from tapping the calls. What's the difference? I'm sure that behavioral advertisers could create a profitable business by listening in. If I tell my best friend I'm pregnant over the phone or by letter, data aggregators could sell that information to companies, which could target me with ads for baby products. But it's a crime for someone to listen in or to open the letter. Yet if I provide the same information in an email or through Skype, data aggregators can collect and sell the information without my consent.

The 2001 *DoubleClick* decision seriously gutted federal protections that were enacted to protect against unauthorized access to computers and digital wiretapping. Yet the need for such protections has increased exponentially since the advent of social networks. The type of information people now transmit over their computers is more intimate and sweeping than when *DoubleClick* was decided a decade ago.

People who want to protect their second selves—their digital doppelgängers— are now turning to state laws that protect privacy or that criminalize unauthorized access to computers as an alternative to the federal acts. In Arkansas, a man who installed software on his wife's computer that copied all her keystrokes, allowing him to obtain his wife's passwords, was found to have violated the Arkansas computer trespass law.[75] In Florida, a woman installed spyware on her husband's computer.[76] At certain intervals, the program captured shots of what the husband saw on his screen, including messages, emails, and websites. The wife was found to have violated the Florida wiretap statute.

Some state laws specifically offer greater protections than the federal laws do. Unlike the federal laws, which don't apply if one party consents to the spying, several states, including Michigan and Washington, require both parties to consent. *In re DoubleClick* might have come out differently under a two-party consent statute. "DoubleClick, for example, has permission from commercial websites to intercept their web communications with their users," says lawyer Jessica Belskis. "However, individual users generally do not consent to the interception of personal information."[77] Belskis asserts that DoubleClick would have been found liable in states with two-party consent laws. She also notes that Google might be liable in those states for scanning the contents of emails for key words to determine which relevant advertisement to display. Gmail subscribers give consent to the scans by agreeing to Google's terms of service, but individuals who exchange emails with Gmail users do not consent to the interception and scanning of their communications.

In April 2011, a California federal judge considered a suit by consumers against NebuAd, which had contracted with internet service providers to install deep packet inspection devices on the networks to monitor and transmit data to NebuAd.[78] The data aggregator had tried to get the suit dismissed, saying that since it wasn't liable under federal law, it couldn't be liable under state law. The judge, though, let the lawsuit go forward under that state's invasion of privacy law and the state's computer crime law.[79] NebuAd went out of business, but other deep packet inspection companies are entering or preparing to enter the U.S. market.[80]

Some lawmakers are trying to enact new laws to protect people, such as a do-not-collect-data list similar to the do-not-call list. But data aggregators are incredibly flush with funds and can fight back strong against individuals who bring lawsuits and lawmakers who suggest laws. When ordinary people sued a data aggregator for putting Flash cookies on their computers without their consent, the defendant's law firm (a powerful law firm of more than a thousand lawyers) accused the plaintiffs' lawyer of operating a "shakedown" operation by filing the suit.[81] A shakedown? The defendants are collecting private information about plaintiffs without their consent, including sensitive health information and financial information, and the plaintiffs are the ones engaged in a shakedown?

When a California lawmaker introduced a law in 2011 that would allow people to opt out of data collection, Facebook, Google, Time Warner Cable, 24/7 RealMedia, the California Chamber of Commerce, and 31 other associations and companies wrote a letter in opposition to the bill. They said, "The measure would negatively affect consumers who have come to expect rich content and services through the Internet, and would make them more vulnerable to security threats."[82] Double Speak again. How does it make me more vulnerable to security threats if I *stop* people from getting unauthorized access to my credit card and Social Security numbers?

The Federal Trade Commission has been the main governmental watchdog protecting consumers from violations of their rights by social networks, data aggregators, and advertisers. The Federal Trade Commission is an independent government agency created in 1914 to prevent unfair competition by businesses and to bust the trusts that had formed when powerful companies worked together to stifle competition in an industry.[83] Later, Congress expanded the FTC's consumer protection powers so that it could take action against "unfair or deceptive acts or practices,"[84] by bringing lawsuits or adopting rules and regulations for specific industries.[85]

If the FTC believes an organization is engaged in an "unfair or deceptive act or practice" or is violating a consumer protection statute, it can issue a complaint setting forth the charges.[86] The organization can elect to settle the charges by signing a consent agreement. If the FTC accepts the agreement, it places the agreement on the record for 30 days for public comment before making a final decision about the charges.[87] If the organization contests the charges, a trial-type proceeding is conducted in front of an Administrative Law Judge.[88]

Over a four-year period from 2004 to 2008, the FTC received 1,230 complaints under the category "company does not provide any opportunity for consumer to opt out of information sharing," 1,678 complaints that the "company fails to honor request to opt out/opt-out mechanism does not work," and 534 complaints that the "company is violating its privacy policy."[89] The agency also received 84 complaints that a "privacy policy is misleading, unclear, or difficult to understand," 555 complaints that a "company does not have adequate security," and 3,265 other complaints of privacy violations.

In the past few years, the FTC has taken on Facebook, Google, and a variety of data aggregators and has forced some changes and collected some damages for consumers. In one case, parents concerned about their children's internet safety bought software from Echometrix to inform them if their children were engaged in inappropriate online behavior. Echometrix's software Sentry Parental Controls monitored and recorded activity on a target computer, including the history of websites visited and chat and instant message conversations.[90] What users of Sentry didn't know was that Echometrix was also creating a database of their information, including excerpts from their children's actual online chat conversations.[91] That information was then sold to data aggregators.

The parents who'd bought the Sentry program had to accept an End User License Agreement (EULA) that, 30 paragraphs from the beginning of the EULA, said: "[Sentry] uses information for the following general purposes: to customize the advertising and content you see, fulfill your requests for products and services, improve our services, contact you, conduct research, and provide anonymous reporting for internal and external clients."[92]

The FTC went after Echometrix for "unfair and deceptive trade practice." Echometrix agreed to ban the use of information collected by Sentry and order the destruction of the collected information.[93] But if Echometrix had included a clearer paragraph in its EULA about Sentry, the FTC likely wouldn't have had a cause of action to pursue Echometrix.

The FTC is now trying to develop regulations for behavioral advertising. "When you're surfing the Internet, you never know who is peering over your shoulder or how many marketers are watching," FTC Commissioner Jon Leibowitz said at a commission hearing on the subject. "People should have dominion over their computers. The current 'don't ask, don't tell' in online tracking and profiling has to end."[94]

The images and information I post on my Facebook page may come back to haunt me. But at least I have a choice about whether to friend my boss or post a photo of myself chugging tequila. I can make careful choices about what I reveal about myself, reflecting on what to reveal about my likes and dislikes, which groups to join, and how much personal information to reveal.

But while I'm busy creating one type of digital self, thousands of decisions are being made about me each day based on my second self—a Lori Andrews of the Web who has been pigeonholed into a marketing subcategory and is being treated accordingly. The law does little to protect me from privacy violations, psychological and economic harms, and the loss of basic rights. Social networks have magnified the problems because we post information on them that is more private and more potentially damaging than the information that is available due to our searches and purchases.

The redlining of the 1960s is now the weblining of the twenty-first century. If we had a Social Network Constitution, we could include a right of privacy that would ban surreptitious data collection. We could include a principle of procedural fairness that would require that we be given advance notice and due process, meaning we would have to opt in to the use of our information, rather than having the information collected secretly or through a process that occurs unless we opt out. And provisions in favor of free speech and against discrimination would prohibit data about our second self from being used inappropriately against us.

The problem with the current situation is that no one has yet created the conceptual framework to determine how we should protect our second selves. This is the very thing that a Social Network Constitution should do. A Constitution for any nation determines what property rights and privacy rights are retained by the citizens. It sets forth the procedures and circumstances under which rights can be denied. So far, in the social network realm, little attention has been given to the rights of individuals. By exploring how other technologies have been greeted by the courts and how people have managed to assert their rights, we can begin the process of creating a Social Network Constitution.

4

Technology and Fundamental Rights

When a young lawyer in Boston married the daughter of a Senator, he was un-prepared for the incessant media attention to their union. After their children were born, paparazzi would snap photos of the babies when the family took walks down the street. Annoyed, he thought about what legal recourse he might have.[1] Were there any legal precedents for a "right to be let alone"?

The lawyer had graduated second in his class at Harvard Law School. He got in touch with his friend, who'd graduated first in the class, and they began a project to analyze how fundamental legal values could be applied to a new technology.[2]

The year was 1889. And the new technology was the portable camera.

Before 1888, when Kodak introduced a portable camera, taking someone's photo was a big deal.[3] A person would get dressed up and go to a studio. Photos were not taken without a person's permission. But the portable camera changed all that.

The two lawyers, Samuel Warren and Louis Brandeis, weren't the only people concerned about the new technology. A newspaper article at the time said:

> Have you seen the Kodak fiend? Well, he has seen you. He caught your expression
> yesterday while you were in recently talking at the Post Office. He has taken
> you at a disadvantage and transfixed your uncouth position and passed it on to
> be laughed at by friend and foe alike. His click is heard on every hand. He is
> merciless and omnipresent and has as little conscience and respect for proprieties
> as the verist hoodlum. What with Kodak fiends and phonographs and electric
> search lights, modern inventive genius is certainly doing its level best to lay us all
> out bare to the gaze of our fellow-men.[4]

Warren and Brandeis began to assess the impact of the portable camera on modern life. They could have suggested that people no longer had a right to be let alone because technologies could now track and record what they did. Instead, they noted that the intrusiveness of technologies made it even more important for people to have control over information about themselves. "The intensity and complexity of life attendant upon advancing civilization have rendered necessary some retreat from the world," they wrote, "so that solitude and privacy have become more essential to the individual; but modern enterprise and invention have, through invasions upon his privacy, subjected him to mental pain and distress, far greater than could be inflicted by mere bodily injury."

Warren and Brandeis turned to fundamental Constitutional values, such as the right to refuse to testify against oneself, and common law principles, such as the "right of determining, ordinarily, to what extent his thoughts, sentiments, and emotions shall be communicated to others." They said that these rights do not depend upon the particular method of expression adopted. "The same protection is accorded to a casual letter or an entry in a diary and to the most valuable poem or essay, to a botch or daub and to a masterpiece."

The Boston lawyers felt this protection was due to a larger fundamental value. "The protection afforded to thoughts, sentiments, and emotions . . . is merely an instance of the enforcement of the more general right of the individual to be let alone," they said. "It is like the right not to be assaulted or beaten, the right not to be imprisoned, the right not to be maliciously prosecuted, the right not to be defamed."

The right of a person to control what was disseminated about him or her is also akin to a property right. That same year, E. L. Godkin, editor of *The Nation*, wrote in *Scribner's Magazine* that one's reputation was "the very first form of individual property, the earliest of individual belongings."[5]

The two Boston lawyers showed how photographs and gossip could harm not only the individual, but society. "Even gossip apparently harmless, when widely and persistently circulated, is potent for evil," they wrote. "It belittles by inverting the relative importance of things, thus dwarfing the thoughts and aspirations of a people." They also took into consideration the proper functioning of a democracy by proclaiming that "Some things all men alike are entitled to keep from popular curiosity, whether in public life or not, while others are only private because the persons concerned have not assumed a position which makes their doings legitimate matters of public investigation."

Their article, "The Right to Privacy," was published in 1890 in *The Harvard Law Review*.[6] Their ideas were incorporated into law through the creation of four distinct legal actions for invasion of privacy: for intruding on someone's seclusion, for publicly disclosing embarrassing facts, for putting a person in a "false light" in the public eye, and for appropriating someone's name or likeness for commercial use. Information about and photos of people can be disseminated if the people have consented or the matter is of legitimate public interest. The fundamental Constitutional right to privacy has additionally been understood to cover a right to make important personal decisions, such as whether to use contraception or whether to homeschool your child.

The mode of analysis of the two Boston lawyers from a century ago has been used to analyze each new technology that has reached the courts. How does it affect the individual and society? How do fundamental legal values help to protect the individual when the technology is used? As each new technology has been adopted—including forensic technologies, medical technologies, and computer technologies—the application of fundamental values has been used to protect, and often expand, people's rights. Sometimes courts, lacking the comprehensive analysis of technology like the one undertaken by Warren and Brandeis, took missteps when they first encountered a technology. But ultimately, the fundamental rights prevailed.

That is, until social networks. When confronted with conflicts in this area, courts have allowed social networks, data aggregators, and third parties using social network information to run roughshod over the rights of individuals.

Why do courts and policy makers fail to apply fundamental democratic principles to social networks? Perhaps it's due to judges' lack of familiarity with how social networks operate. Perhaps it's because technologies are entering our lives at an unprecedented rate, making it seem that they are impossible to control. Or perhaps it's because no one has bothered to review what has happened when other technologies were introduced that challenged the ideals of consumer protection, due process, freedom of expression, privacy, and control that are the foundation of the U.S. Constitution.

Other technologies that courts have encountered share many characteristics with social networks. They, too, involve people's control of information and decisions about themselves and their activities. With respect to every other technology, courts have not hesitated to apply Constitutional principles to protect users' rights. Understanding those decisions can assist in the drafting of a Social Network Constitution.

When Charles Katz entered a public phone booth in 1965, he never imagined that cops would tap the phone line. The cops charged him with placing illegal bets—and he protested that they'd infringed the Fourth Amendment limits on governmental intrusion into a person's private life. The trial judge said that wiretapping didn't violate the Fourth Amendment because the Founding Fathers drafted that Constitutional provision to honor people's privacy in their homes. In this case, the police hadn't trespassed into his home. In fact, there had even been a Supreme Court decision on the matter, back in 1928, when cops had used earlier wiretap technology to learn that someone was violating prohibition.

In that earlier case, *Olmstead v. United States*, the five-justice majority of the U.S. Supreme Court had held that a bootlegger's privacy hadn't been invaded and he hadn't been forced to incriminate himself because, although police had recorded the calls he was making from his home, the wiretap equipment had been placed on phone lines outside his home.[7] Writing for the dissent was none other than the former Boston lawyer Louis Brandeis, who was now a Supreme Court justice. He argued that fundamental values had to be applied to new technologies. He noted that when the Constitution was adopted, "force and violence"—torture and breaking into people's houses—were the only ways that the government had to obtain private information about people. The Constitution protected against force and violence. But, said Brandeis, "discovery and invention have made it possible for the government, by means far more effective than stretching upon the rack, to obtain disclosure in court of what is whispered in the closet. . . . The progress of science in furnishing the government with means of espionage is not likely to stop with wiretapping. Ways may some day be developed by which the government, without removing papers from secret drawers, can reproduce them in court, and by which it will be enabled to expose to a jury the most intimate occurrences of the home." According to Brandeis, the Constitution's fundamental value of privacy and the right not to incriminate yourself needed to be applied not only to "what has been, but of what may be."

Forty years after the *Olmstead* decision, when Charles Katz's case was appealed to the U.S. Supreme Court, the majority of the justices applied Brandeis's logic. Even though Charles Katz was using a public phone booth, the Court said that the Constitutional right of privacy "protects people, not places." What a person seeks to preserve as private, even in a public place, may be Constitutionally protected.

The Supreme Court protected Katz's privacy by enunciating a legal test that is still used today: Did the person have an "expectation of privacy," and was that an expectation that society was willing to protect? As a result, police need to get a warrant, based on probable cause, before they tap someone's phone.

The march of law enforcement technology continued, and in 2001 a new forensic technology reached the court. A federal agent suspected Danny Kyllo of growing marijuana.[8] Since growing pot indoors required high-intensity lamps, the agent sat in a car across from the house and used an Agema Thermovision 210 thermal imager to scan Kyllo's home. The scan showed that the roof over the garage and a side wall of the home were relatively hot compared to the rest of the home and substantially warmer than neighboring homes in the triplex. The agent concluded that Kyllo was growing pot and convinced a judge to allow him to search Kyollo's home. The agent found pot, and Kyllo was convicted on a drug charge. Because the thermal scanner did not physically intrude on the house and did not show any private human activities, the trial court said that it hadn't infringed Kyllo's Constitutional rights.

The appellate court, too, held that Kyllo had shown no subjective expectation of privacy because he had made no attempt to conceal the heat escaping from his home, and "even if he had, there was no objectively reasonable expectation of privacy because the imager 'did not expose any intimate details of Kyllo's life,' only 'amorphous "hot spots" on the roof and exterior wall.'"[9]

When the U.S. Supreme Court took the case, it reversed Kyllo's conviction. The opinion protecting Kyllo against this new technology was written not by a liberal justice following the approach of Brandeis, but by one of the conservatives. "It would be foolish to contend that the degree of privacy secured to citizens by the Fourth Amendment has been entirely unaffected by the advance of technology," wrote Justice Antonin Scalia. "Where, as here, the Government uses a device that is not in general public use, to explore details of the home that would previously have been unknowable without physical intrusion, the surveillance is a 'search' and is presumptively unreasonable without a warrant."

Advances in medical technologies, too, have impeded people's control of their lives and even of their deaths. Life-extending technologies such as respirators, feeding tubes, and resuscitation equipment were initially used on people even when they didn't consent. The government and medical facilities asserted that they had the authority to extend the patient's life, even over the patient's objection. But ultimately fundamental Constitutional values prevailed against the onslaught of technologies.

Now people's right to refuse life-sustaining treatment is clearly recognized. They can even issue advance proclamations, such as living wills, about avoiding such technologies if they slip into a coma and can't make their wishes known. In explaining the importance of this right under the Constitution, Supreme Court Justice John Paul Stevens pointed out that a person has a right to control her image, "an interest in being remembered for how she lived rather than how she died."[10] A woman who was a vibrant athlete might decide that she'd rather have her friends and family members remember her healthy, energetic self. She might decide to fill out an advance directive saying she'd rather not be kept alive if she was in an irreversible coma, so that those vibrant memories of her would not be overwritten by ones of a wasting body connected to machines. Individuals should be able to control the image other people have of them.

Contemporary medical technologies, such as genetic testing, have also raised disputes. When genetic testing became possible, people were tested without their knowledge or consent. Doctors and researchers would use blood that people had given to labs for routine cholesterol or pregnancy tests and perform additional testing, without the person's consent, for everything from breast cancer to Alzheimer's disease. The argument was, what's the harm? The person has already been pricked; the additional tests involved no additional intervention. And even if the blood was collected anew—as in a forensic DNA test—blood tests were safe and noninvasive.

But then employers and insurers started discriminating against healthy people based on their genetic predisposition to future disease. With certain genetic mutations, for example, some women had a higher risk of developing breast cancer than other women. Even with those mutations, half the women would not develop breast cancer. Some women didn't want to know whether they had the mutations or not. They said they would then feel like they had a time bomb ticking away inside them. But employers and insurers wanted that information to make their decisions. There were no legal limits on what could be done with that information.

During routine physicals, an employer in California had the company doctor surreptitiously test the female employees to see if they were pregnant and the African-American employees to see if they carried the sickle cell anemia gene mutation. The results were not disclosed to the employees, but they were put into the employees' personnel files.

When the existence of the files leaked, the employees sued. The trial court dismissed the case, saying that the test was a modest intrusion, no more than what people usually undergo in a physical. But the appellate court held that genes contain personal information that is protected by the fundamental right to privacy. "One can think of few subject areas more personal and more likely to implicate privacy interests than that of one's . . . genetic make-up," wrote the court.[11] Since then, Congress has passed a law specifically prohibiting employers and insurers from discriminating against people based on the results of genetic tests.[12] People's fundamental rights include the right not to have genetic information generated about them or used against them.

Even computer technologies that collect data about people have been subject to a fundamental rights analysis in the pre-social networks era. When Judge Robert Bork was nominated for the U.S. Supreme Court in 1987, Michael Dolan, a Washington, D.C., newspaper reporter, attempted to discredit him by publishing his video store rental records. In today's world, Judge Bork's choices seem tame: British movies, Bond movies, costume dramas.[13] The reporter was disappointed not to see legal movies such as *12 Angry Men* or *To Kill a Mockingbird*. Instead, Judge Bork had rented "only one truly court-related tape": *The Star Chamber*.

Bork did not get the Supreme Court nomination. But the publication of his video rentals did get the attention of Congress. "It is nobody's business what Oliver North or Robert Bork or Griffin Bell or Pat Leahy watch on television or read or think about when they are home," said Senator Pat Leahy. "In an era of interactive television cables, the growth of computer checking and check-out counters, of security systems and telephones, all lodged together in computers, it would be relatively easy at some point to give a profile of a person and tell what they buy in a store, what kind of food they like, what sort of television programs they watch, who are some of the people they telephone. . . . I think that is wrong. I think that really is Big Brother, and I think it is something that we have to guard against."[14]

Senator Paul Simon agreed. "There is no denying that the computer age has revolutionized our world. Over the past 20 years we have seen remarkable changes in the way each one of us goes about our lives. Our children learn through computers. We bank by machine. We watch movies in our living rooms. These technological innovations are exciting and as a nation we should be proud of the accomplishments we have made. Yet, as we continue to move ahead, we must protect time honored values that are so central to this society, particularly our right to privacy. The advent of the computer means not only that we can be more efficient than ever before, but that we have the ability to be more intrusive than ever before. Every day Americans are forced to provide to businesses and others personal information without having any control over where that information goes. . . . These records are a window into our loves, likes, and dislikes."[15]

Senator Leahy also noted that the trail of information generated by every transaction that is now recorded and stored in sophisticated record-keeping systems is a new, subtler, and more pervasive form of surveillance. "These 'information pools' create privacy interests that directly affect the ability of people to express their opinions, to join in association with others, and to enjoy the freedom and independence that the Constitution was established to safeguard."[16]

The legislators applied the fundamental Constitutional right to privacy and passed a law forbidding disclosure of people's video rental records (or, in this day and age, what they watch on Netflix). The bill prohibits video stores from disclosing "personally identifiable information"—information that links the customer or patron to particular materials or services. In the event of an unauthorized disclosure, an individual may bring a civil action for damages.

"There's a gut feeling that people ought to be able to read books and watch films without the whole world knowing," said Representative Al McCandless, a co-sponsor of the bill. "Books and films are the intellectual vitamins that fuel the growth of individual thought. The whole process of intellectual growth is one of privacy—of quiet, and reflection. This intimate process should be protected from the disruptive intrusion of a roving eye."[17]

Now courts are facing a new set of technologies—social networks and data aggregators. One would think that judges would seek to protect people since these technologies raise all the issues that courts have previously found problematic. People's privacy is being invaded when their data is collected. They can't exercise control over their image or reputation. Crucial judgments about them—such as whether they will be hired or retain custody of their children—are made based on photos they are tagged in and titles of movies they like.

When confronted with cases about social networks, though, courts have failed to apply fundamental values and previous legal precedents. After visiting her hometown of Coalinga, California (population 19,000),[18] a University of California at Berkeley student, Cynthia Moreno, posted an "An Ode to Coalinga" on Myspace, using just her first name. The ode, which she took down six days later, began, "The older I get, the more I realize how much I despise Coalinga."[19]

The Coalinga High School principal made the ode available to be published in a local newspaper, identifying its author. Cynthia's parents heard a rumor about the impending publication and contacted the editor of the *Coalinga Record*, and explained the damage that republishing the ode would cause.[20] The editor then purportedly promised that she would not publish the ode in the *Coalinga Record*.[21]

But the ode was published. And when the newspaper appeared, the community reacted with an uproar. Cynthia received death threats, as did the rest of her family—her parents and younger siblings—who still lived in Coalinga.[22] A car drove up to the family home, and someone fired a gun at the house, hitting the family dog and narrowly missing Cynthia's baby brother. The parents fled from their home out of fear. Cynthia tried to calm things down, asking the newspaper to publish a letter from her straightening things out. The newspaper editor refused and instead published hate mail directed to Cynthia. At the local high school, the teachers showed Cynthia's ode to their classes, leading to more hate mail being sent to Cynthia. People boycotted her parents' business. Her parents, Mexican immigrant farmworkers who had worked their way up, had to default on the property loan for the modest trucking company they owned in Coalinga.

Cynthia could not have anticipated the devastation that would come from her rant, which echoed that of many of her other college friends who'd been thrilled to escape their small towns to pursue their educational dreams. On Myspace, a few friends had commented on her posting. But she hadn't intended for it to be read by the citizens of Coalinga.[23] And since she hadn't spoken disparagingly of any particular person in the town, she was shocked by the threats of violence. She talked to friends at the law school at Berkeley, who told her that if there was one thing that was true in publishing law, it was that no one could publish something you'd written without your permission.

She and her parents took the matter to the school board to ask for an apology from the principal. When the school board said it couldn't help her, they appealed that decision to the state board of education and, ultimately, to the U.S. Department of Education in Washington, D.C. There, a helpful employee told her to get a lawyer and sue the school district.

But social networks were in their infancy, and dozens of lawyers turned Cynthia down. Finally, an attorney agreed to handle the case, suing the newspaper for intentional infliction of emotional distress and invasion of privacy.

The court dismissed her privacy claim entirely. When she appealed the ruling, a California appellate court held that "no reasonable person would have had an expectation of privacy" for a posting on a social network.[24]

And when the intentional infliction of emotional distress claim later went to trial, a jury held that even though the principal had behaved outrageously in submitting the ode for publication without her permission, all the horrific threats were a reaction to the Myspace posting rather than the newspaper publication.[25] That's like saying that smoking a cigarette is akin to throwing one into a tank of gasoline. If the ode had appeared only on Myspace, her parents would still have had their home and business. It was its publication in the *Coalinga Record* that turned Cynthia and her family into targets.

On the opposite coast, a New York judge also refused to protect privacy online: "In this day of wide dissemination of thoughts and messages through transmissions which are vulnerable to interception and readable by unintended parties, armed with software, spyware, viruses and cookies spreading capacity; the concept of internet privacy is a fallacy upon which no one should rely. It is today's reality that a reasonable expectation of internet privacy is lost, upon your affirmative keystroke."[26]

Courts are abdicating their responsibility to protect individuals when they reason that you can't expect privacy on the internet since nothing on the Web is safe from someone with hacking skills or a data aggregation agenda. That's like saying we shouldn't have laws against Peeping Toms, since if you have a house with windows, anyone can look in. Or that if rape is common in a certain neighborhood, a woman can't legitimately have an "expectation to be free from rape." By defining the expectation of privacy based on the fact that any digital information is accessible by someone, the courts have failed to recognize the important social value of privacy and the fact that people often believe that social networks are private spaces.

It's almost as if Janlori Goldman, a privacy expert, had social networks in mind when she testified more than 20 years ago at the hearings triggered by the release of Bork's video rentals. "The new technologies not only foster more intrusive data collection, but make possible increased demands for personal, sensitive information," she said. "Private commercial interests want personal information to better advertise their products. The government is interested in sensitive information to enhance political surveillance. And, the intelligence community may be looking at reading lists to protect our national security. The danger here is that a watched society is a conformist society, in which individuals are chilled in their pursuit of ideas and their willingness to experiment with ideas outside of the mainstream. . . . New technologies enable people to receive and exchange ideas differently than they did at the time the Bill of Rights was drafted. Personal papers once stored in our homes are now held by others with whom we do business. Transactional information may be easily stored and accessed. Records of our reading and viewing histories are now maintained by libraries, and cable television and video companies. The computer makes possible the instant assembly of this information."[27]

Goldman advocated the application of fundamental Constitutional rights to personal online information, but the courts in social network decisions have done just the opposite. The ubiquity of social networks seems to have discouraged policy makers from trying to regulate them. But with previous technologies, the fact they were ubiquitous ultimately weighed in favor of assuring protection, instead of assuming that nothing could be done. When the Supreme Court protected Charles Katz from being wiretapped in a phone booth, it said, "To read the Constitution more narrowly is to ignore the vital role that the public telephone has come to play in private communication."

Perhaps we are just at the early stages of the use of this technology. After all, it was less than a decade ago that Mark Zuckerberg launched Facebook from his room at Harvard. It took nearly 40 years for the fundamental Constitutional rights approach to be applied to wiretapping. But we can't wait 40 years. By then everything we've ever transmitted on our computers will be part of a database like Spokeo or PeekYou. Schools, employers, mortgage brokers, credit card companies, and courts in custody cases will have made decisions against us based on information from social networks.

The Warren and Brandeis article not only created a legal framework that still applies today to safeguard people's privacy, it also established a method for judging new technologies. The authors analyzed how fundamental values inherent in the U.S. Constitution and common law provide a basis to make judgments about new technologies. They also assessed how the new technology affected individuals, institutions, and the larger society. Warren and Brandeis did not suggest that individuals adapt to each new technology, but instead advocated that society assure that technology was employed in a way that was consistent with fundamental historical societal values.

When Brandeis was appointed to the U.S. Supreme Court 26 years after his privacy article appeared, he continued to champion the application of Constitutional values to modern technologies. He also wrote about the nature of a Constitution. "Time works changes, brings into existence new conditions and purposes. Therefore a principle, to be vital, must be capable of wider application than the mischief which gave it birth. This is peculiarly true of Constitutions. They are not ephemeral enactments, designed to meet passing occasions. They are, to use the words of Chief Justice Marshall, 'designed to approach immortality as nearly as human institutions can approach it.' The future is their care, and provision for events of good and bad tendencies of which no prophecy can be made. In the application of a Constitution, therefore, our contemplation cannot be only of what has been but of what may be."[28]

A wealth of technologies has been developed since the Warren and Brandeis article appeared. Some have initially been used in ways that ran roughshod over people's individual rights. Courts have been called in to sort out the disputes and their decisions have been all over the map. But eventually, the Constitutional principles favoring self-determination, privacy, due process, and personal control have prevailed. With past technologies, it was anything goes at first—then courts and legislators began to put constraints into place, preserving fundamental values. For the past two centuries, courts have examined new technologies (from the portable camera to high-tech medical technologies) and applied the fundamental principles underlying the Constitution to uphold and even expand people's rights.

Social networks have come to play vital roles in our public and private lives. But this march of technology does not have to trample the concepts of consumer protection, restraints on governmental power, and protection of individual rights. We could create a Social Network Constitution applying fundamental values if we launched a concerted effort to do so. For example, the principle of due process advance notice would require that people be told what would be done with their information when they enter a social network. The principle of control would allow the users of social networks to prevent third-party use of their information, preferably through a mechanism that prevents further disclosure unless the user has specifically opted in. Right now, some social networks and data aggregators allow people who read the fine print to opt out of further disclosure. But if you don't even know that a data aggregator like Spokeo exists, how can you be expected to opt out?

"Facebook describes itself as a 'social utility,' as if it's a twenty-first century phone company," says Eli Pariser, author of *The Filter Bubble: What the Internet Is Hiding from You*.[29] "But when users protest Facebook's constantly shifting and eroding privacy policy, Zuckerberg often shrugs it off with the caveat emptor posture that if you don't want to use Facebook, you don't have to. It's hard to imagine a major phone company getting away with saying, 'We're going to publish your phone conversations for anyone to hear—and if you don't like it, just don't use the phone.'"[30]

Today, when a Senator's son-in-law walks down the street, his photo could be snapped by a nearby pedestrian's smartphone. Through the PeekYou technology, his image could be instantaneously scanned and linked to existing online databases. The stranger, curious about this chance passerby on the street, would be shown a profile of the Senator's son-in-law—what he does for a living, whether he uses internet dating sites, his latest tweets, his latest Yahoo! searches, his Facebook page. The profile may not provide an accurate understanding of the actual person on the street. Yet it transforms him from a mere form on the street to a person lugging around his intimate life experiences. Accomplishments and desires become as easily attributable as eye color and height.

A century has passed since the Kodak fiend, and rather than capturing a snapshot of a moment, the technologies of today seek to expose a person's entire life story.

Today's technologies refuse to let people alone, attempting to reconstruct people's past while authoring their future. The need to protect individual rights to lead a full and social life in the face of intrusive technology has never been greater. The personal and social freedoms necessary for human flourishing today are much the same as they were at the time the U.S. Constitution was adopted. A Social Network Constitution, capable of shielding people against the distorting lens of public scrutiny and the misuse of social network information, promises to align the latest technologies with some of society's oldest values.

5

The Right to Connect

In 2011, young Egyptians turned to Facebook, Twitter, and YouTube to coordinate and broadcast their plans to protest.[1] On January 15, 2011, Wael Ghonim, creator of the Facebook group "We are All Khaled Said," posted an event page encouraging Egyptians to protest on January 25, 2011.[2] Wael originally created the group in response to the murder of businessman Khaled Said by Egyptian police in June 2010. Prior to his death, Khaled had inadvertently acquired footage that appeared to implicate police officers in corruption.[3] In the video, the police officers on a drug bust seemed to be dividing up the seized drugs and money among themselves. One of the police officers said in the video, "Now it's time for a vacation."[4] The video had shown up on Khaled's computer—his relatives posit that the officers were sharing the video via Bluetooth in an internet café below Khaled's apartment.[5] He posted the video online and was killed by two of the police officers in the video several weeks later.[6]

Wael announced on Facebook that the protest would take place on Police Day, a national holiday in Egypt.[7] The link to the event was then shared on Twitter by 21-year-old Alyouka (@alya1989262), who tweeted, "http://on.fb.me/fBoJWT over 16000 of us are taking to the streets on #jan25! join us: http://on.fb.me/fQosDi #egypt #tunisia #revolution."[8] A few days later, 26-year-old Asmaa Mahfouz posted a YouTube video on Facebook, announcing, "We want to go down to Tahrir Square on January 25th. . . . We'll go down and demand our rights, our fundamental human rights."[9] The video soon went viral.[10] By January 25, 2011, the event page on Facebook had over 95,000 confirmed attendees.[11]

The virtual turned real that day as tens of thousands of people gathered in Tahrir Square.[12] The power of social networks to seed a revolution became clear. But President Hosni Mubarak was not about to let a bunch of kids with laptops and mobile phones threaten his 30-year regime.[13] Two days after the initial Tahrir Square protest, when the protesters in Egypt tried to use the internet, they discovered that they could not access Facebook, Twitter, or any other social networking site.[14] Mubarak had pulled the plug on the nation's internet.

At first, outsiders thought the Egyptian government had hit a physical kill switch, perhaps turning off equipment in Cairo's Internet Exchange Point,[15] a facility where internet service providers (ISPs) converged to interconnect, trade traffic, and connect to providers outside the country.[16] However, Renesys, a U.S. internet monitoring firm, observed that, contrary to appearances, Egypt's internet shutdown was "not an instantaneous event on the front end."[17] A timeline of events showed that four of Egypt's five internet service providers (ISPs) all removed themselves from the internet within one to six minutes of each other. So instead of a big red button scenario, Renesys theorized that Egypt's government called the four providers and each shut down its part of the Egyptian internet. The ISPs likely withdrew their Border Gateway Protocols, the mechanism by which ISPs announce their customers' internet protocol (IP) addresses to other ISPs to help make connections. As a result, their customers' IP addresses were made invisible to other ISPs and therefore the rest of the world. For four days, only the Noor Group internet service provider remained active, possibly because it hosted the Egyptian stock exchange. However, on January 31, 2011, the Noor Group also went offline.[18]

Even though the Mubarak regime had shut down the internet and mobile phone service, people continued to take to the streets.[19] The protesters found ingenious ways to post at least a trickle of information on social networks. Some used fax machines and dial-up modems to connect to internet service providers outside Egypt at the cost of a long-distance phone call.[20] Others, particularly those living near the border, were able to get around the internet blockage by piggybacking on cell phone service from neighboring countries, such as Israel.

When direct reports from their relatives were blocked out, Egyptians in other parts of the world flocked to an island called Egypt in the virtual world of Second Life. Users' avatars carried signs supporting the Egyptian protesters, Arab music played loudly, and audio feeds brought in real time discussions of Egyptian policy in Arabic.[21] People immersed in this digital town square could feel energy akin to that in Tahrir Square. Egyptians in other countries also used Second Life to send chat instructions on how to covertly communicate with people in Egypt.[22]

If anything, the internet shutdown only added to the protesters' resolve. And the Egyptian government found out how costly it is to shut down the internet. On the morning of February 2, 2011, Egypt's government restored the internet.[23] The Organisation for Economic Co-operation and Development (OECD) estimated that the five-day internet shutdown cost Egypt's ISPs at least $90 million in lost revenues.[24] This figure did not include the secondary economic impacts that resulted from a loss of business in other sectors such as e-commerce, tourism, IT, and call centers.

The protests continued and on February 11, 2011, President Mubarak resigned from office. Following his resignation, expressions of gratitude for the social network appeared in Cairo as graffiti, saying "Thank you Facebook."[25] And when a baby girl was born soon after, her Egyptian father named her "Facebook" to commemorate the role of the social networking site in the January 2011 revolution.[26]

Three weeks later, Libya disappeared from the internet.[27] Unlike Egypt where the government likely told ISPs to shut down their servers, Libya's Border Gateway

Protocol routes were still visible and open.[28] There was just no traffic. As James Cowie, co-founder and chief technology officer of Renesys, told *International Business Times*, "It's like a post-apocalyptic scenario where the roads are there, there just isn't any traffic."[29]

Libya's primary internet provider, Libya Telecom & Technology, is state-owned and maintains a monopoly over Libya's international internet gateway.[30] Most likely, the Libyan government ordered Libya Telecom's operators to throttle the rate at which its servers accepted data down to zero, where no data could get in or out.[31] So although Libya's servers were technically still online, the fact that they could not transmit or receive information rendered Libya's internet useless.

In the United States at that same time, Congress was considering the need for a kill switch for the U.S. internet—not to silence dissidents, but to respond to cyberterrorism. Prior to the Egyptian uprising, Senators Joseph Lieberman, Susan Collins, and Tom Carper had introduced a law, the Protecting Cyberspace as a National Asset Act of 2010, which would establish a federal Director of Cyberspace Policy who could be given the authority to issue an emergency shutdown directive to internet service providers.[32]

But killing the internet in the United States would not be as easy as in Egypt or Libya, where there are just a few internet service providers.[33] The Egyptian government most likely caused the blackout with just a few phone calls because there are only five major internet service providers in Egypt. And Libya has just one primary internet service provider. The United States has 2,000 to 4,000 ISPs, many of which are privately owned.[34] To replicate the Egyptian internet kill switch model, the U.S. government would have to make a thousand times as many phone calls, with many of the call recipients potentially ignoring the order to shut down.[35]

But the U.S. government could put a major dent in internet access by targeting the largest internet service providers, the exchange points, and wireless providers all at once. The top five ISPs—AT&T, Comcast, Road Runner, Verizon, and America Online[36]—account for half the U.S. internet market, while the top ten cover 70%.[37]

The Department of Homeland Security is working on another possibility—a plan to add digital signatures to the internet traffic routing information used by ISPs to help connect with one another, which in turn would allow ISPs and enterprises to authenticate the information and prevent hackers from redirecting traffic. Adding digital signatures to routing data would allow the agency in charge of issuing digital signatures to refuse to route traffic from an unauthenticated block of IP addresses from a domestic ISP or possibly even another country's ISP.[38] Such a mechanism would change the basic internet philosophy of openness to one where the federal government holds the keys to stopping traffic to and from entire portions of the United States and perhaps the world.[39]

Private companies are already offering technologies to U.S. security agencies and foreign governments to monitor the content of suspected dissidents' online transmissions. In 2006, Steve Bannerman, marketing vice president of Narus, a

California-based internet filtering and surveillance company, told Wired.com about Narus's powerful internet inspection technology products. "Anything that comes through (an internet protocol network), we can record," Bannerman said. "We can reconstruct all of their e-mails along with attachments, see what web pages they clicked on, we can reconstruct their (voice over internet protocol) calls."[40]

Later in 2006, Mark Klein, a longtime AT&T technician, started to see what he considered curious connections between AT&T and the National Security Agency.[41] Klein kept quiet about the connections until news broke that President George W. Bush had authorized the NSA to eavesdrop (without warrants) on Americans with suspected ties to Al Qaeda. In the subsequent class-action lawsuit filed by the Electronic Frontier Foundation against AT&T for helping the NSA invade customers' privacy, it was revealed that Narus had manufactured the equipment used by the NSA to monitor information from telephone and email communications through AT&T's network.[42] The court case was dismissed in 2009 on the grounds that the company had immunity under a federal anti-terrorism law.[43] In 2010, Narus was acquired by the large defense contractor Boeing to operate as a subsidiary of Boeing's defense business.[44]

The possibility of the U.S. government having the legal and technological power to stop access to the Web or monitor transmissions caused concern among internet users and sparked talk of building a citizens' internet of wireless devices that could evade potential government constraints. As the opposition to a kill switch grew, President Obama went on YouTube's World View during the Egyptian shutdown to disavow the possibility of a similar blackout in America. Amazingly, the president spoke of social networks as if they were already part of a Constitutional mandate. President Obama asserted, "There are certain core values that we believe in as Americans, that we believe are universal—freedom of speech, freedom of expression, people being able to use social networking or any other mechanisms to communicate with each other and express their concerns."[45]

Social networks provide astonishing power to breathe life into fundamental rights. They can enhance freedom of association, freedom of the press, and freedom of expression. They can be used for everything from taking down a government to announcing the birth of a child. Facebook, Twitter, and Myspace have become central to how we live, how we work, and how we play. They play a key role in shaping who we are as individuals and as a society. Should our Social Network Constitution contain a right to connect? If so, how would such a right be justified, what would it consist of, and would it have any limits?

Most fundamental rights are directed towards individuals—such as the right to liberty, the right to privacy, and freedom of speech. But the right to connect would also protect institutions that offer connections, much as the individual right of freedom of speech has its counterpart in freedom of the press.

The original idea of freedom of the press came about in the 1600s in response to laws that chilled the dissemination of information and opinions, especially about the government. In England until the end of the seventeenth century, no publication was allowed without a prior state-issued license. Public criticism of the government was not only illegal, but punishable by death.

These measures were criticized by the philosopher John Milton, who felt that the marketplace of ideas—where citizens had access to information that could challenge or support their ideas—was crucial to democracy. Citizens needed to be able to weigh different viewpoints in order to make sound political decisions. John Stuart Mill, whose philosophical treatise *On Liberty* provided foundational rules for modern democracy, similarly advocated that an individual's right of free speech be nearly absolute—limited only when it harmed another person.

When the American colonists set about to express their fundamental values, they adopted measures to ensure Milton's marketplace of ideas. They rejected the British approach of requiring licenses in advance of publication and also rejected penalties for criticizing the government. In fact, they felt that political speech, especially if it criticized the government, should be the most highly protected form of expression. After all, the colonists' dissatisfaction with the British government had just led them to a revolution. The result was the First Amendment to the U.S. Constitution: "Congress shall make no law . . . abridging the freedom of speech, or of the press."

Today, the U.S. Department of State touts American freedom of the press around the world. Through a publication available in multiple languages, the State Department notes that "Although a cherished right of the people, freedom of the press is different from other liberties of the people in that it is both individual and institutional. It applies not just to a single person's right to publish ideas, but also to the right of print and broadcast media to express political views and to cover and publish news."[46]

The State Department admits that "Accurate information will not always come directly from the government, but may be offered by an independent source, and the maintenance of freedom and democracy depends upon the total independence and fearlessness of such sources."

The right to freedom of the press includes the right to publish anonymously—a right that had not been possible under the British licensing laws. The right to anonymity encourages speakers to enter into the marketplace of ideas by alleviating concerns of future retaliation, economic harm, social ostracism, or invasion of privacy. But the U.S. Supreme Court has indicated that anonymity offers more than just a way to escape persecution for one's ideas.[47] Alienated speakers might resort to anonymity to deliver their message to a larger audience without the message being prejudged.[48]

Anonymity was central to the political papers that provided impetus behind the United States Constitution. Although the U.S. Constitution was written in 1787, at least nine of the 13 original colonies were required to ratify it for it to take effect.

In 1788, three of the Founding Fathers, Alexander Hamilton, James Madison, and John Jay, published a key document, *The Federalist Papers: A Collection of Papers in Favor of the New Constitution*. The publication did not list their names, but instead was printed under the pseudonym Publius, a reference to Publius Valerius Publicola, who helped overthrow the Roman monarchy and became a Roman consul in 509 B.C. The essays in *The Federalist Papers* were convincing. States came on board, and the Constitution went into effect in 1789.

Legal battles over the scope of freedom of the press and the protection of anonymity continue to this day. The U.S. Supreme Court has recognized that freedom of the press necessarily includes a right to gather news as well.[49] And, although early court cases protected the right to anonymity in the context of political expression,[50] courts in some instances have protected anonymous speech that's not directly about the government, such as anonymous criticism of a corporation.[51]

Like John Stuart Mill's idea that freedom of expression can be limited when it does harm to someone else, the rights of freedom of expression and of anonymity are limited in certain narrow circumstances. The protection of anonymity is weighed against other social interests,[52] such as the right of a person to pursue legal claims against a wrongdoer.[53]

Social networks and the internet have created an explosion in the magnitude of information available and the means of connection. People can now communicate with citizens all over the country and all over the world. Most people now get their news from the Web. Through Google Books, they can gain access to more than 15 million books,[54] an estimated 12% of the world's books.[55] All of a sudden, the marketplace of ideas is not the size of a neighborhood store but the digital equivalent of Minnesota's Mall of America.

New technologies also make it easier to express yourself anonymously. Easily and inexpensively, a person who is posting can use a proxy server (such as WiTopia, Cryptohippie, or Identity Cloaker) to conceal the IP address of the computer from which the post was made.[56] If the particular website requires registration, the poster could create a new email address, again using a proxy server. Anonymity software such as Tor[57] relays communications through a series of networks to conceal the original IP address.[58] This software protects primarily against "traffic analysis"—determining which websites a particular computer visits.

Social networks, web access, and anonymizing software provide the foundation for an unparalleled richness of political discussion on the Web. At the highest levels of the U.S. government, the right to use social networks—and the internet more generally—is taking its place as a fundamental value. "The rights of individuals to express their views freely, petition their leaders, worship according to their beliefs—these rights are universal, whether they are exercised in a public square or on an individual blog," Secretary of State Hillary Clinton said. "The freedoms to assemble and associate also apply in cyberspace."[59]

Since the Founding Fathers' vision was forward looking, the fundamental values expressed in the U.S. Constitution are broad enough to embrace freedom of expression on the internet. The U.S. Supreme Court has recognized the importance of digital discourse. "Through the use of chat rooms, any person with a phone line can become a town crier with a voice that resonates farther than it could from any soapbox," the Court observed in *Reno v. ACLU*. "Through the use of Web pages, mail exploders, and newsgroups, the same individual can become a pamphleteer."[60]

Anonymity on the internet is being protected as well. Noting that the internet "allows people from all over the world to exchange ideas and information freely and in 'real-time,'" a federal trial court in Washington in the case *Doe v. 2TheMart.com* implied a special premium on internet anonymity, given its ability to facilitate "the rich, diverse, and far ranging exchange of ideas."[61] The court refused to compel disclosure of the identities of posters on an investment site who had alleged that the company 2TheMart.com had lied and defrauded customers.[62] Yet courts have also recognized that anonymity can facilitate wrongdoing—shielded by a pseudonym, a wrongdoer may engage in defamation, copyright infringement, or other wrongful activities without fearing any redress.[63] And these are problems that need to be addressed.

Similar to the U.S. Constitution, the United Nations' Universal Declaration of Human Rights states: "Everyone has the right to freedom of opinion and expression; this right includes freedom to hold opinions without interference, and impart information and ideas through any media regardless of frontiers." Digital freedom of expression, including access to social networks, enables the propagation of other fundamental rights valued by the United Nations, such as the rights to freedom of association and assembly, the right to education, and the right to take part in cultural life.[64] Yet the organization Reporters Without Borders indicates that one-third of the world's people live in countries that do not have freedom of the press. For people in those countries, social networks can open up new channels for information which the governments traditionally denied them.

Some countries have declared the rights to connect to the internet and to access social networks as fundamental human rights. In June 2009, France's highest court declared that "'the free communication of ideas and opinion is one of the most precious rights of man' . . . and given the generalized development of public online communication services and the importance . . . for the participation in democracy and the expression of ideas and opinions, this right implies freedom to access such services."[65]

In contrast to traditional newspapers, the instantaneous transmission of Facebook and Twitter can allow protesters to get their message out before their communications can be shut down. And, while police might be able to attack a bricks-and-mortar building housing a newspaper or television station, anonymity can make a Facebook or Twitter source untraceable.

Estonia now guarantees its citizens a Constitutional right to connect and to obtain information over the internet.[66] People are entitled to have an internet access point close to them available at a reasonable cost[67]—or even free if they can't afford it.[68] As a result, Estonia is now one of the most digitally advanced countries. It also has one of the highest rankings in terms of freedom of the press. According to the 2010 Index of Press Freedom compiled by Reporters Without Borders, the United States ranks 20th of the 178 countries ranked on issues of press freedom. The Northern European countries ranked at the top of the list. Estonia was 9th, France 44th, Israel 86th, Egypt 127th, and Iran 175th.[69]

These rankings consider not just what a country says but what it does. Merely acknowledging freedom of speech or of the press is not enough. Prior to the Tahrir Square incident, Egypt already had a provision in its Constitution providing, "Liberty of the press, printing, publication and mass media shall be guaranteed."[70] But the Egyptian government required a licensing system for newspapers and, through its role as a co-owner of the three largest newspapers, controlled their distribution and content.[71] Reporters and bloggers who criticized the government were harassed and imprisoned. From January 2009 to March 2009 alone, 57 journalists in Egypt were taken to court for anti-government statements. Bloggers were subject to imprisonment for denigrating the president, covering labor disputes, and commenting on religious issues.[72]

The United Nations Human Rights Council has assessed what the fundamental freedom of expression and opinion means for the Web. A 2011 report, commissioned by the council and prepared by human rights attorney Frank La Rue of Guatemala, determined that freedom of expression should include (1) mostly unrestricted access to online content and (2) access to the physical and technical infrastructures required to access the online content in the first place. The report said that countries should prioritize "facilitating access to the Internet for all individuals, with as little restriction to online content as possible."[73] It also called upon countries to ensure a person's ability to use the internet anonymously. The report also warns that governments should not be allowed to monitor or collect information about individuals' internet communication, since that would "impede the free flow of information and ideas online."[74] The report advocates that internet freedom be limited only in rare instances, such as to protect national security.

Rights related to social networks and the internet were also the focus of the 2011 G8 Summit. The leaders of the G8 nations summarized the issues addressed at their summit in a Declaration: "For citizens, the Internet is a unique information and education tool, and thus helps to promote freedom, democracy and human rights. The Internet facilitates new forms of business and promotes efficiency, competitiveness, and economic growth. Governments, the private sector, users, and other stakeholders all have a role to play in creating an environment in which the Internet can flourish in a balanced manner. In Deauville in 2011, for the first time at Leaders' level, we agreed, in the presence of some leaders of the Internet economy, on a number of key principles, including freedom, respect for privacy and intellectual property, multi-stakeholder governance, cyber-security, and protection from crime, that underpin a strong and flourishing Internet."[75]

Should there be any limits on our Social Network Constitution's right to connect? The United Nations report recommends that, even in cases where the government or a private entity may be thought to have a reason to restrict the right to connect, such a restriction must be carried out only if it is necessary and is the least restrictive way to protect the rights of others or to further important societal goals such as national security.

But we need to be wary of a national security exception, since it might swallow up the right to connect. After all, in the 1600s in England, criticizing the government was considered to be detrimental to national security and it was subject to punishment, even if it was true. In fact, the rule in the United States is quite the contrary. The United States was founded on the idea that a marketplace of ideas is necessary for a democracy. Political speech is protected, sometimes even if it is false. If a public figure is defamed or falsely accused in the press, the publication cannot be sued unless they knew in advance that the material was untrue and acted with malice in printing it.[76]

When WikiLeaks published 251,287 cables from more than 250 U.S. embassies around the world, the disclosure was criticized as potentially embarrassing government officials, possibly damaging intelligence efforts, and taking freedom of the press too far. But such challenging, embarrassing, and privacy-invading possibilities were exactly what the Founding Fathers had in mind. They were willing to support freedom of the press even when it was turned on them.

When Founding Father Alexander Hamilton found his extramarital affair with Maria Reynolds exposed in a pamphlet that included letters they had written to each other, he did not lose faith in freedom of the press. Instead, he published his own pamphlet admitting the affair. Despite being targeted by the press, Hamilton remained a staunch supporter of its freedom. Hamilton later defended Harry Croswell, the editor of the New York–based Federalist paper *The Wasp*, against charges of treasonous libel against President Thomas Jefferson.[77] Hamilton argued, "The liberty of the press consists, in my idea, in publishing the truth from good motives and for justifiable ends, though it reflect on the government, on magistrates, or individuals."[78] Jefferson, the subject of the libel, too, felt that freedom of expression was crucial. "They fill their newspapers with falsehoods, calumnies and audacities," Jefferson told a friend. "I shall protect them in their right of lying and calumniating."[79]

Political speech might be chilled, though, by schemes such as the Department of Homeland Security's introduction of digital tags, which identify who's communicating by deep packet inspection snooping into individuals' transmissions. Yet there are reasons to be concerned about cybersecurity. In September 2010 in Iran, a computer virus designed to disrupt power grids and other industrial facilities infected the computers of the country's nuclear power plant at Natanz.[80] The virus, Stuxnet, is the first known computer virus designed to target physical infrastructures such as power stations.[81] It first recorded readings of normal operations at the nuclear plant and then played those readings back to plant operators while the virus caused the centrifuges to spin abnormally.[82] The virus set back Iran's nuclear program.

Senator Lieberman referred to Stuxnet to justify his previously introduced bill to provide a mechanism to shut down the U.S. internet. But an internet kill switch would not have prevented Stuxnet from reprogramming the centrifuge motors and interfering with Iran's nuclear facilities. This is because Stuxnet targeted industrial systems not usually connected to the internet to prevent such attacks from happening in the first place. This suggests that the virus gained access to the Windows-based computers through other means, such as via USB ports. The virus must have been planted manually.[83]

According to internet security expert Jonathan Zittrain, "it's not clear that government intervention would make any difference" in the event of a massive computer-virus or malware attack. He points out that the ISPs would already be doing everything they could to counter the attack and that the government would not "have any comparative advantage in understanding the situation better than the Internet engineers themselves."[84] And by creating more centralization, Homeland Security's tagging system might actually make the system more vulnerable to anyone with access to the tags or who is targeting the tags. Giving the government enhanced abilities to monitor the internet will not offer sufficient security benefits to offset the chill on freedom of expression and the assault on anonymity that it will entail.

Technologies for tracking online messages or intercepting them undermine the freedom to connect. In the wake of the Egyptian protests, it was revealed that Narus, the American company, had sold its deep packet inspection technology to Egypt's state-controlled ISP, Telecom Egypt, which may have facilitated spying on dissidents and other citizens.

During the February 10, 2011, House Committee of Foreign Affairs hearing entitled "Recent Developments in Egypt and Lebanon," Congressman Chris Smith noted that Narus had sold deep packet inspection technology to Egypt, which may have been used by the Egyptian government to "identify, track down, and harass or detain" dissident journalists in Egypt.[85] Congressman Bill Keating questioned Deputy Secretary of State James Steinberg on what the Department is doing to ensure that American internet inspection technologies sold to other countries "are not an obstacle to human rights at best, or a tool of violence at worst."[86] Comparing social media tools to guns, Keating pledged to introduce a bill "that would provide a national strategy to prevent . . . American technology from being used by human rights abusers."[87] Keating said the legislation he would create for internet inspection technologies, such as deep packet inspection, would have "the same safeguards—such as end user monitoring agreements—that we do when we sell weapons abroad."[88] When the United States sells weapons abroad, it places various restrictions on what the purchasing country can do with the weapons as part of the end use monitoring agreement.[89] Such restrictions can include who and what the weapons cannot be used against. But in focusing on how other countries use our tracking technologies, Congress may be ignoring the use of the technologies in the United States itself.

What if the right to connect is subject to a private limitation, not a governmental one, such as when someone is denied the right to connect or the right to post certain information due to intellectual property concerns? Even then, the balance should weigh in favor of the right to connect. France adopted a law to authorize the tracking of internet users to determine if anyone had pirated online copyrighted material and to automatically cut off internet access to those who continued to download illegally after two warnings.[90] France's highest court, the Constitutional Council, held that the law violated "the right of any person to exercise his right to express himself and communicate freely, in particular from his own home." The law unconstitutionally breached privacy by enabling the government to monitor people's internet activity[91] and violated the freedom to connect by cutting off further internet use by the alleged offender without a judicial hearing. Such an approach also runs afoul of the U.N. report finding that cutting off an individual's internet access based on intellectual property rights is "disproportionate."

Copyright laws have also been used to directly silence consumer speech. Social networks and the internet in general have empowered people by allowing them to share opinions and to post reviews of places, services, and professionals. Although there are as many doctors as restaurants in the United States, restaurant reviews are abundant, while reviews of doctors are not, despite the fact that the choice of a doctor is much more important to one's life than the choice of a restaurant. The internet is changing that, as patients start posting online reviews of their doctors. Now some doctors are forcing patients to sign contracts that say that the doctor owns the copyright in any review that a patient writes about him. That way, if the review is critical of the doctor, he or she can use the Digital Media Copyright Act to have the review taken down, just as Sony Pictures could require a takedown of an illegally posted version of its newest film on the ground that it violated Sony's copyright.

To address this issue, Santa Clara University's High Tech Law Institute and the Samuelson Law, Technology & Public Policy Clinic at the University of California Berkeley School of Law have joined forces to create a website, Doctored Reviews, to help patients, doctors, and review-website operators understand the problems created by the "anti-review contracts."[92] Such clear anti-consumer and anti-expression maneuvers by doctors would not be permitted under our Social Network Constitution's right to connect.

In 1787, Thomas Jefferson proclaimed, "The basis of our government being the opinion of the people, the very first object should be to keep that right; and were it left to me to decide whether we should have a government without newspapers or newspapers without a government, I should not hesitate a moment to prefer the latter."[93] With Facebook and Myspace crossing national borders and attracting more citizens than almost any other country, social networks are now like newspapers without governments. But how should social networks themselves be regulated?

Social networks can invigorate freedom of the press, freedom of association, and freedom of expression. Yet even in the liberal democracies, where so many of the social networks were created, the right to connect is threatened. The expansion of governmental powers to fight terrorism, the move to disable internet access for those who violate copyright laws, and other technological and policy developments could inappropriately interfere with fundamental rights.

Our Social Network Constitution should start with a right to connect. Such a right promotes the freedom of digital expression and enables many other fundamental freedoms, such as those guaranteed by the Universal Declaration of Human Rights. Our right to connect should not be abridged by concerns about intellectual property infringement nor about vague notions of national security. The right to connect should not be diminished by any surreptitious online monitoring, including the use of technologies such as deep packet inspection, cookies, and web scrapers. These tracking technologies not only violate users' right to privacy but also hamper the activities that fuel the online marketplace of ideas.[94] Confidence in the security of the internet diminishes; the free flow of information ebbs.

Our Social Network Constitution should also guarantee a right to anonymity. Anonymous use of the internet not only helps protect fundamental rights, but also ensures the personal safety of individuals expressing their opinions and participating in democracy. Internet anonymity empowers ordinary people to participate in the public forum, thus equalizing power differences and serving as a democratizing force in public discourse, notes Lyrissa Barnett Lidsky, professor of law at the University of Florida Levin College of Law.[95] But at the same time, anonymity on the internet may magnify particular harms, such as defamation, by reaching a large audience, allowing for easy replication (as through reposting), and making it difficult to correct misstatements.[96]

Randi Zuckerberg, then the marketing director for Facebook, declared in 2011 that "anonymity on the internet has to go away."[97] Eric Schmidt, when he was the CEO of Google, also called anonymity "dangerous" and suggested that governments will require identification.[98] Doing away with anonymity would help Facebook's and Google's bottom lines (since they use people's personal information to sell ads). But forbidding anonymity could lead to grievous harm. The ability to identify people who post on the internet led to the imprisonment of protesters in Egypt and can facilitate the persecution of individuals and groups.

Data aggregation about identifiable people can quickly turn into a means to target them. Even information originally collected for benign purposes can turn deadly. In *Delete: The Virtue of Forgetting in the Digital Age*, Viktor Mayer-Schönberger describes how the Dutch government in the 1930s created a registration system to keep better track of its citizens. The population registry listed each citizen's "name, birth date, address, religion, and other personal information" for the purpose of better facilitating government administration and welfare planning. However when the Nazis invaded the Netherlands during World War II, they seized possession of the registry and used it to go after the Dutch citizens who were Jewish or gypsy. The repurposed registry was so comprehensive, notes Mayer-Schönberger, that the Nazis were able to identify and ruthlessly murder more than 70% of the Jewish population in the Netherlands, as opposed to 40% in Belgium and 25% in France.[99]

Of course, unrestricted rights can also lead to great harm. In cases such as child pornography, hate speech, and inflammatory incitements to commit genocide, restricting the right to connect is not a disproportionate violation. But even if a government or private entity has a socially valued reason to restrict the right to connect, such a restriction must be carried out only if it is the least restrictive way to protect the rights of others.

Because of social networks, the marketplace of ideas is mushrooming, but so are the concerns for the people who post and the people about whom they post. Photos and postings from people's Facebook and Myspace pages are being held against them in the workplace, in the legal system, and at school. What people are posting and showing on social networks may invade other people's privacy or spread lies about them. As with any Constitution, tough decisions will need to be made about whether the right to connect should be limited in some way by other rights. In the offline world, for example, freedom of the press must often be balanced in criminal cases against another Constitutional right, that of the defendant to a fair trial. Our touchstone should be the one introduced by John Stuart Mill of robust freedom of expression unless it creates a substantial harm to a private individual. Such a formulation would put less value on concerns raised by the government about potential security problems or raised by businesses about potential intellectual property issues and would instead focus on furthering connectivity, political discourse, and social interchange.

How do we further the fundamental freedom to connect without abridging other fundamental rights? That question will be key as we create and assess the other provisions of our Social Network Constitution.

6

Freedom of Speech

Eighteen-year-old Nick Emmett was an ideal student. A senior at Kentlake High School in Washington, he had a 3.95 grade point average, was co-captain of the basketball team, and had never gotten in trouble at school or at home. He took his class assignments seriously. In English class, his teacher asked the students to write fake obituaries of themselves. Later, at home, Nick continued the enterprise and made up fake, tongue-in-cheek obituaries about two of his friends. He posted them on a fake web page he entitled "Unofficial Kentlake High Home Page." His friends loved it. Other students asked him to write their obits. And he modified the web page so that his friends could vote on whose obits he should write next.

Then all hell broke loose.

In the post-Columbine era, any association between "death" and "student" was incendiary. Local television news characterized his website as a "hit list." Nick immediately deleted the website. But the next day, he was called into the principal's office and expelled for intimidation, harassment, disruption to the educational process, and violation of Kent School District copyright (for using school images on his "unofficial" website). The expulsion was winnowed back to a five-day suspension, with the punishment enhanced by forbidding him to play in a crucial basketball game.[1]

In South Carolina, a firefighter-paramedic, Jason Brown, used a simple web program, Xtranormal, to create an animated video that he posted on Facebook.[2] In the three-minute video, a character who looks like an animated Lego fireman responds to a 911 call placed by a doctor in a hospital ER. The doctor wants the fireman to take the patient to another hospital. He says the patient has been there for eight hours and has had a cold for ten days. When the fireman asks what care the patient has received, the doctor says "Nothing, really," explaining that "the big game was on TV and we didn't want to miss the commercials." The fireman is incredulous and tells the doctor how to give the patient some basic care.

The next day, Jason's boss, the director of the county fire rescue service, fired him because of the video. In his termination letter, the boss said that Jason "displayed poor judgment in producing a derogatory video. . . . This video has created an embarrassing situation for this department, our public image and the cooperative relationship we enjoy with Colleton Medical Center."[3] Yet the animation had not named the doctor, the EMT, or the hospital.

The U.S. Constitution generally protects freedom of expression—and both the student and the firefighter believed they were undertaking a protected creative activity. The right to free speech is well known. The Supreme Court has held that high school students' free speech rights include publishing school newspapers and wearing black armbands to protest war. State statutes protect the rights of employees who are whistleblowers or otherwise point out flaws in their workplace. A federal law protects workers who criticize their employment conditions in order to encourage collective action for change.

But, in contrast to the protection of offline speech, people have been penalized for exercising their rights of expression on Facebook or other social networks. Top students have been suspended, expelled, or turned over to the juvenile courts[4] when they posted a criticism of a teacher or coach—or even when they made satirical comments on the poor physical condition of their school and its resources.[5] Employees have found themselves out of a job when they swear in a post or criticize their boss.

The stakes are high. In many high schools, students' grades can be dropped a full grade point (for example, from an A to a B for every class) for each ten days of school missed.[6] If a student is suspended for ten days due to a social network posting, he or she can go from passing to failing or from the Ivy League to Podunk U. based on that suspension.

That's what happened to Brandon Beussink, a junior at a Missouri high school, who, at home, created a fake website of his school that was critical of his teachers and the administrators.[7] At home he showed it to a buddy, who, back at school, showed it to a teacher. She went to the principal, who suspended Brandon. His grades were docked, which meant he was failing all of his classes.

In another case, Justin Layshock used obscene, vulgar, and profane language in creating an unflattering fake Myspace profile of his principal.[8] A gifted student, Justin was enrolled in advanced placement classes and had won awards in interscholastic academic competitions. As punishment, the principal pulled him out of his classes and stuck him into a program designed for students who could not function in a classroom.

In Pennsylvania, it was a teacher whose Facebook frustrations came back to haunt her. Associate Professor Gloria Gadsden of East Stroudsburg University was suspended for status updates that she posted on Facebook. One post read, "Had a good day today, didn't want to kill even one student. :-) Now Friday was a different story."[9] A second post read: "Does anyone know where I can find a very discreet hitman? Yeah. It's been that kind of day."[10] Even though Gloria had initially set up her page as private and hadn't friended any students, her posts had become public without her knowledge after Facebook changed its privacy settings in December 2009.[11] After a student reported the comments to school administrators,[12] the school suspended the professor for a month and forced her to undergo a psychological evaluation before returning to work.[13] According to Marilyn Wells, interim provost and vice president for academic affairs, "Given the climate of security concerns in academia, the university has an obligation to take all threats seriously and act accordingly."[14]

Some employees try to be more careful. Two unhappy restaurant workers created a Myspace page that was password protected, meaning that only individuals who had been invited could view the page or comment on the content.[15] The page was designed to provide a place for employees of the Houston's restaurant at Riverside Square in Hackensack, New Jersey, to "vent about any BS we deal with at work without any outside eyes spying in on us. This group is entirely private, and can only be joined by invitation."[16] When their boss put pressure on one of the workers to give him the password, he read the postings and promptly fired the employees who'd created the website.

What rights of free speech should social network users have, particularly when they are posting their remarks in their free time at home, rather than at school or work? Such postings are not without consequences. Teachers or other students might feel intimidated by postings. Customers might shun a business because of negative comments by employees.

The issue is particularly tricky with underage students. We don't let students drive until they are 16 or drink until they are 21 because of the dangers of such behaviors. But we encourage children to acquire computer skills when they are as young as age four. The middle school student who taunts a classmate on Facebook or the high school student who defames a coach might end up the defendant of a lawsuit or suspended from school, either of which means that their college prospects have gone up in smoke.

Allowing a 13 year old access to Myspace or Facebook can be the equivalent of handing them the car keys in terms of the damage they can do both to other people and to themselves. An eighth grader in a Bethlehem, Pennsylvania, school created a Teacher Sux website that included reasons why Mrs. Fulmer, his algebra teacher, should die and asked for $20.00 "to help pay for the hitman."[17] The site had a devastating impact on Fulmer. She was afraid to go outside of her house. The principal called the FBI, which tracked down the student, who was then expelled. In another Pennsylvania case, an eighth-grade student created a fake Myspace profile for her principal that implied he was a pedophile.[18] If it had been taken seriously, the principal's job would have been on the line.

But in a culture that values freedom of speech, should students and employees be punished for the words they type on the Web? Nick Emmett and his parents took the fake obituary case to court to quash his suspension so that he could attend class and play basketball.[19] Federal District Court Judge John C. Coughenour said that "students do not abandon their right to expression at the schoolhouse gates."[20]

The judge quoted from the seminal U.S. Supreme Court case on the matter,[21] *Tinker v. Des Moines*, which had upheld the First Amendment rights of students to wear black armbands protesting the Vietnam War, even though their school district had banned the armbands. The judge noted that prohibition of students' expressive conduct is justifiable only if the conduct "would materially and substantially interfere with the requirements of appropriate discipline in the operation of the school."[22]

Nick's school had persuasively argued that schools needed to exercise caution in light of school shootings and that internet postings "can be an early indication of a student's violent inclinations, and can spread those beliefs quickly to like-minded or susceptible people."[23] But the judge pointed out that the school had presented "no evidence that the mock obituaries and voting on this web site were intended to threaten anyone, did actually threaten anyone, or manifested any violent tendencies whatsoever."[24] Since the school had infringed Nick's freedom of speech, the judge ordered that the school readmit him and wipe the suspension from his record.

Similarly, Brandon Beussink's postings about his teacher were found to be protected by the First Amendment. A federal judge in Missouri, Rodney W. Sippel, declared, "Disliking or being upset by the content of a student's speech is not an acceptable justification for limiting student speech." The judge noted that "One of the core functions of free speech is to invite dispute." Quoting a Supreme Court decision, he said that speech "may indeed best serve its high purpose when it induces a condition of unrest, creates dissatisfaction with conditions as they are, or even stirs people to anger. Speech is often provocative and challenging."[25]

"Indeed," wrote the judge, "it is provocative and challenging speech, like Beussink's, which is most in need of the protections of the First Amendment." He held that schools cannot penalize students for what they write on websites or social networks unless it is "speech within the school that substantially interferes with school discipline."[26]

Nor have courts been comfortable upholding criminal convictions of students who speak derogatorily of school officials on social networks. In Indiana, a middle schooler posted vulgar comments on a private profile that another student had created in the name of the principal of her previous school. She then created a public Myspace group page on which she directed an expletive to the principal. But the Indiana Supreme Court said the teen wasn't guilty of harassment for her private post since she hadn't even intended that the teacher see the site, let alone be harassed by it. As to the public post, harassment requires that the action be taken with no other intent than to harass. The court gave her the benefit of the doubt, indicating that she might not have realized that the group was public. As to the intent to harass, the court found it "even more plausible that A.B., then fourteen years old, merely intended to amuse and gain approval or notoriety from her friends, and/or to generally vent anger for her personal grievances."[27] The court reversed her conviction.

The eighth grader who created the "Teacher Sux" website did not fare as well. The Pennsylvania Supreme Court said his threat (soliciting $20 for a hitman to knock off his algebra teacher) was not protected by the First Amendment because it had caused "actual and substantial disruption" at school by causing the teacher emotional distress, requiring the hiring of a substitute teacher, and providing a hot topic of conversation among students. But the court went further, characterizing the website as on-campus speech even though it had been created at home. Because the website was aimed at the specific audience of students and others at the school, rather than a random audience, "it was inevitable that the contents of the web site would pass from students to teachers, inspiring circulation of the web page on school property."[28] Under this standard, any off-campus speech that addresses school-related issues could be regarded as on-campus speech, entitled to less Constitutional protection. A dissenting judge at the trial level had disagreed, saying that the boy had never intended to harm anyone and had posted a disclaimer on his website that teachers were not supposed to look at it. That judge pointed out that the school district had never perceived the boy's actions as a real threat. The school hadn't sent him to a psychologist to determine if he was a danger, for example. What's more, said the judge, "This type of sick humor can be found in some of today's popular television programs, such as *South Park*."[29]

In the school cases, the critical inquiry was whether the postings disrupted school activities. In a case where a student had created a fake web page with ribald attacks on his coach (including aspersions about the man's penis size), the court said, "We cannot accept, without more, that the childish and boorish antics of a minor could impair the administrators' abilities to discipline students and maintain control."[30] In another case, the purported "disruption" was that students decorated the offender's locker to show support for her posting.[31] To allow that to be considered a disruption would mean that a student who won a sporting event or appeared on *American Idol* could similarly be disciplined if someone decorated her locker.

Employees are sometimes no more circumspect than eighth graders in their posts—and often have fewer rights than the kids who attend public school. When 22-year-old Connor Riley was offered a job at Cisco, she tweeted, "Cisco just offered me a job! Now I have to weigh the utility of a fatty paycheck against the daily commute to San Jose and hating the work." Someone at the company read the tweet, and Connor's job offer was withdrawn. "We here at Cisco are versed in the web" was the response Connor received to her tweet.[32]

A 16-year-old girl was shocked when she was fired after a coworker saw that the girl had written on Facebook that her job was boring, even though she hadn't mentioned the company's name.[33] (She wrote, "first day at work. Omg!! So dull!!" and "all i do is shred holepunch n scan paper!!!")[34] In response, a union official criticized the firing and drew a parallel between Facebook and social conversations: "Most employers wouldn't dream of following their staff down to the pub to see if they were sounding off about work to their friends."[35]

Employees who work for sports teams often let their team spirit get into the way of their employment prospects. In 2010, 24-year-old Andrew Kurtz was fired from his job of dressing as a pierogi and running in the pierogi races during the fifth inning stretch of Pittsburgh Pirates home games. He'd posted a comment on his Facebook page disagreeing with his employers' decision to extend certain managers' contracts: "Coonelly extended the contracts of Russell and Huntington through the 2011 season. That means a 19-straight losing streak. Way to go Pirates."[36]

The federal agency overseeing workplace disputes—the National Labor Relations Board—is concerned that the law is not keeping pace with social networks. The National Labor Relations Board chair, Wilma B. Liebman, says, "American labor law, enacted when the prototypical workplace was the factory, and the rotary telephone was 'the last word in desktop technology,' increasingly appears out of sync with changing workplace realities."[37]

Employees currently don't fare as well as public school students in their First Amendment claims. While students have a right to attend public school, most employees are subject to the "at will doctrine," which allows bosses to fire people without any reason at all. There's an outer limit to the employers' power, though— a firing must comply with standard policies of the employer, and an employer can't uncover the social network posting in a way that violates other state and federal laws.

An employee's freedom of speech is also protected if he or she is a public employee commenting on an issue of public importance or a whistleblower revealing dangerous or corrupt practices at work. But that is a much narrower scope of protection than that given public school students, whose freedom of expression is limited only if it causes disruption to the school or poses a threat to someone. Workers do have some protection under the National Labor Relations Act—if their postings on social networks fall within their legal right to engage in "concerted activities for the purpose of collective bargaining or other mutual aid or protection."[38]

Jason Brown, the fireman with the ER animation, might be able to claim that his video revealed important public information or was part of concerted activities to improve the lot of EMTs. Emergency medical services are one of our most important public services. Any abuse of 911 resources is a cause for concern, since it could prevent someone with a legitimate medical emergency from receiving appropriate care. After he was fired, Jason Brown told a local news station that the video was about "just general things that go on in the day-to-day business of us running calls within any fire department, any EMS."[39] If there are doctors like the one in Brown's video who are misusing 911 resources, it's important that hospitals become aware of it and take the necessary action to correct this behavior.

The right to engage in concerted activities protected a Connecticut emergency medical technician. When her boss didn't let her bring in a union rep to help defend her at a hearing about a customer's complaint, Dawnmarie Souza retaliated on her Facebook page. She called her boss a "scumbag" and "dick"[40] and used the company's shorthand for a psychiatric patient—a 17—to say this about her boss: "love how the company allows a 17 to become a supervisor."[41] Her boss fired her, citing a company policy that bars employees from depicting the company "in any way" on Facebook or other social media sites and prohibits employees from making "disparaging or discriminatory comments when discussing the company or the employee's superiors and co-workers."[42]

Dawnmarie's case seems even more problematic than Jason Brown's. She was saying negative things about a real person, while he had created a general animation that did not identify anyone. Yet Dawnmarie's firing was reversed. In November 2010, the National Labor Relations Board considered her case. Lafe Solomon, the board's acting general counsel, said, "This is a fairly straightforward case under the National Labor Relations Act—whether it takes place on Facebook or at the water cooler, it was employees talking jointly about working conditions, in this case about their supervisor, and they have a right to do that."[43]

The National Labor Relations Board also got involved when Deborah Zabarenko, an environmental reporter and head of the Newspaper Guild at Thomson Reuters, was admonished by her employer for a Twitter post. After a supervisor suggested employees post about how to improve Reuters as a workplace, Zabarenko responded to a company Twitter address, "One way to make this the best place to work is to deal honestly with Guild members."[44] The next day, she received a call from the bureau chief, who told her that it was against Reuters' policy for employees to say anything that might hurt Reuters' reputation. Zabarenko said she "felt kind of threatened" and "thought it was some kind of intimidation."[45] According to the NLRB, Reuters' actions had infringed the reporter's right to engage in concerted activity to improve working conditions.

But when the Houston restaurant employees with the password-protected Myspace page sued, claiming that their termination had violated their First Amendment rights, the court rejected their claim because their employer was not a government agency. The court said that even if their employer had been a public institution, the employees didn't meet the standard of showing that their words related to a matter of public concern.[46] The employees were not union members and thus could not turn to a collective bargaining agreement for support.

But a jury did offer the restaurant employees some relief. Since they had tried to keep the Myspace page secret, their boss was guilty of violating both the federal and state Stored Communications Acts for his unauthorized access to their account (by pressuring another employee for the password).[47] Brian Pietrylo and Doreen Marino, the employees who'd set up the Myspace page, were awarded damages of $2,500 and $903, respectively.

And some workers can seek a legal remedy if the company's own handbook creates a procedure for termination that has been violated. In the Pittsburgh Pirates case, the national attention directed at Kurtz's firing led to him being reinstated in his part-time job as one of 18 employees who dress as pierogies. "We hired him back because he was not terminated in accordance with our Human Resources procedures," said team spokesman Brian Warecki. "While his conduct was not in accordance with company policy, it was not subject to termination."[48]

As we create our Social Network Constitution, should there be any limits to the right to free speech? We can begin to explore potential limits by assessing instances in which the U.S. Constitution's First Amendment rights have been outweighed by other policy concerns.

Most famously, free speech can be abridged when it creates an imminent, unwarranted risk of harm. The renowned jurist U.S. Supreme Court Justice Oliver Wendell Holmes stated in a 1919 case, "The most stringent protection of free speech would not protect a man in falsely shouting fire in a theater and causing a panic."[49] That idea of imminent harm outweighing free speech was put to the test in 1971 when *The New York Times* and *The Washington Post* published the Pentagon Papers, which were the WikiLeaks of their time. The papers, a top secret Pentagon analysis of the history of the involvement of the United States in Vietnam, showed that President Lyndon Johnson's administration had systematically lied to Congress and the American public. When the papers were leaked to the press by an insider, the U.S. government tried to prevent their publication. The Supreme Court held that the First Amendment protected their publication—and reiterated that the government can enjoin publications only if there is serious, imminent harm, such as if a newspaper published the location of our troopships at war, making them a target for the enemy.

When Israeli soldiers gained the right to use social network sites, it was accompanied by an extensive campaign to warn soldiers not to disclose military information online. Military bases in Israel were peppered with posters that showed mock Facebook friend requests from enemies of the state: Iranian President Mahmud Ahmadinejad, Syrian President Bashar al-Assad, and the Lebanese Hezbollah leader Hassan Nasrallah. Below their photos was the question, "You think that everyone is your friend?"[50]

But the lure of Facebook and its centrality to modern life overpowered even those warnings. When Israeli soldiers from an elite artillery corps[51] were deployed to the West Bank to capture militants suspected of planning attacks against Israel, one of the soldiers posted the following: "On Wednesday we clean up Qatanah, and on Thursday, God willing, we come home."[52] The rest of his Facebook status update provided the time and place of the planned raid, as well as the name of the soldier's unit.[53] As a result, the Israel Defense Forces called off the West Bank raid and court-martialed the soldier.[54] This example falls squarely within the troopship example given by the U.S. Supreme Court of speech that can be prohibited or penalized because of the possibility it will lead to imminent harm.

In the era of social networks, where so many people feel entitled to be the lead of their own movie, some employees' postings do cross the line of inciting imminent harm or falsely defaming someone. Other postings seem to go to the heart of a profession or risk invading the privacy of patients, clients, or customers.

A soldier's job necessarily involves maintaining secrecy about missions. Similarly, secrecy about patients and clients is central to certain professions. The promise of confidentiality by a health care provider, says one court, "is as much an express warranty as the advertisement of a commercial entrepreneur."[55] Thus it hardly seems shocking or inappropriate that five California nurses were fired for discussing patients' cases on Facebook[56] nor that a 48-year-old doctor, Alexandra Thran, was fired from Westerly Hospital in Rhode Island, reprimanded by the state medical board, and fined $500 after posting about a patient online.[57] Dr. Thran's Facebook page featured some of her emergency room experiences, and, although she did not include the patient's name, others were still able to identify the patient based on Thran's description of the injuries.[58] Lawyers, too, have been fired due to breaching clients' confidentiality, including a public defender who posted about clients on her blog, referring to them by their first names, a derivative of their first names, or their jail identification numbers.[59]

Photos can breach a patient or client's right to confidentiality even more dramatically. Thus it makes sense that a nurse from Southern General Hospital in Glasgow was suspended when she used her cell phone to take photos of patients during operations and posted them on Facebook.[60]

In New York City, EMT Mark Musarella used his BlackBerry to take pictures of 26-year-old murder victim Caroline Wimmer's strangled corpse and put them on Facebook. Caroline's parents, Martha and Ronald Wimmer, filed suit against Facebook, EMT Musarella, his employer Richmond University Medical Center, Fire Commissioner Salvatore J. Cassano, the New York Fire Department, and the owners of Caroline's apartment building. Musarella was fired.[61] He also faced criminal charges and pled guilty to misconduct and disorderly conduct, avoiding a possible one-year jail sentence by agreeing to do 200 hours of community service and never again to work as an EMT.[62]

Social network postings may also undermine the integrity of entire professional institutions. Amanda Tatro, a student in the mortuary-science program at the University of Minnesota, posted status updates about her cadaver, whom she named Bernie, on Facebook. Tatro posted, "Amanda Beth Tatro gets to play, I mean dissect, Bernie today. Lets see if I can have a lab void of reprimanding and having my scalpel taken away. Perhaps if I just hide it in my sleeve. . . ." In another update, Tatro wrote that she was "looking forward to Monday's embalming therapy as well as a rumored opportunity to aspirate. Give me room, lots of aggression to be taken out with a trocar." And when Tatro realized that she would no longer be dissecting Bernie, she posted a farewell, "I wish to accompany him to the retort [the chamber for cremation]. Now where will I go or who will I hang with when I need to gather my sanity? Bye, bye Bernie. Lock of hair in my pocket."

Tatro's privacy settings allowed her updates to be viewed by her Facebook "friends" and "friends of friends," which included hundreds of people. A fellow mortuary-science student reported the posts to program administrators, and Tatro received a failing grade in her anatomy laboratory class, was required to take a clinical ethics class and undergo psychiatric evaluation, and was placed on academic probation. When Tatro challenged the sanctions in court as a violation of her right to free speech, the court observed that Tatro's posts had "presented substantial concerns about the integrity of the anatomy-bequest program."[63] Her posts had weakened the trust of potential donor families, jeopardizing the survival of an institution heavily dependent on the public's perception and goodwill. In professions that touch upon deep cultural sensitivities, posts like Tatro's can easily shake public confidence—making people less likely to donate blood or organs or increasing the emotional angst of those who've lost a loved one.

But what rule should govern the person posting the information who isn't a professional with an independent duty not to blurt? A Brooklyn food delivery man got mad at the small tips he was receiving and he started taking photos of the receipts and posting them—along with customers' names and addresses—on his blog.[64]

The delivery man was fired and that seems right. Among other things, the postings violated the right of the customers to privacy of place, control over the dissemination of their address, and privacy of personal information about them.

Similar concerns would be raised by creditors who attempt to shame debtors by using social networks. For almost a century, cases have established that certain publications about debts can violate a person's right to privacy. In the first such case, in Kentucky in 1926, a garage mechanic posted a large sign saying that the town veterinarian owed him $49.67. The court held that the right of privacy included "the right of a person to be free from unwarranted publicity, or the right to live without unwarranted interference by the public about matters with which the public is not necessarily concerned."[65]

What if the clients or customers disparaged are not named but are just described as part of a group? If an employee's posts indicate resentment towards a particular group, an employer may legitimately be concerned that the employee might discriminate against members of that group.

When students in an Arkansas school initiated a campaign to wear purple on October 20, 2010, to honor gay teens who had committed suicide, a school board member, Clint McCance, posted a response on Facebook. "Seriously they want me to wear purple because five queers killed themselves," he posted. "The only way im wearin it for them is if they all commit suicide. I cant believe the people of this world have gotten this stupid. We are honoring the fact that they sinned and killed thereselves because of their sin. REALLY PEOPLE."[66] He later apologized, noting that his remarks were "very ignorant" and that he did "not wish death on anyone." Although the Arkansas education commissioner denounced the remarks, McCance could not be fired because the position is elected. McCance resigned to spare the school district bad press saying he would consider running again in the future.[67] But McCance's derogatory comments about the gay community suggest an intolerance that could undercut his ability to properly perform his school board duties, which include funding of school groups and activities.

In February 2011, pro-union demonstrators congregated inside the Wisconsin capitol building to protest a bill that would eliminate the collective bargaining rights of public employees.[68] Responding to news that riot police might clear the protesters out of the capitol building, Indiana Deputy Attorney General Jeffrey Cox tweeted that the police should "use live ammunition" when attempting to disperse the protesters.[69] Previously, Cox had made offensive statements via Twitter and on his blog, referring to one Indianapolis resident as "a black teenage thug who was (deservedly) beaten up" by city police.[70] In this case, Cox was not revealing confidential information about a client, but he was posting remarks directed to a group who might enter the legal system. As a representative of the legal arm of the state government, should Cox be held to have a duty to follow legal standards of using the legal system rather than violence to solve problems? After he was fired, the Indiana Attorney General's Office released a statement saying "Civility and courtesy toward all members of the public are very important to the Indiana Attorney General's Office. We respect individuals' First Amendment right to express their personal views on private online forums, but as public servants we are held by the public to a higher standard, and we should strive for civility."[71]

What if the postings disparaged all customers or clients, rather than just naming a few or a particular group? At that point, the postings seem more like general ventings about a job—which should be protected. Such postings would indicate dissatisfaction with a job, rather than something problematic because it signals an imminent risk, a breach of confidentiality, or discriminatory treatment. But it's a tough call.

The issue of an employee's disparaging posts is a particularly tricky one for police departments. Almost one-third of law enforcement agencies have dealt with "negative attention" due to their employees' on-duty or off-duty use of social media, according to a survey of 728 U.S. law enforcement agencies undertaken by the International Association of Police Chiefs.[72] Cops' bravado and their disdain for criminals may be what draws them to the profession in the first place—and may be what allows them to put their life on the line every day. But when cops express those traits online, their posts may be used by police departments to suspend them or by criminal defendants to challenge their credibility.

Trey Economidy was an Albuquerque cop whose Facebook profile described his occupation as "human waste disposal."[73] His profile was private, but a change in Facebook policy made the identity of his employer public.[74] While on duty, Economidy was involved in the fatal shooting of a suspect. Afterward, a local TV station disclosed his profile. Economidy was given desk duty, and the Albuquerque Police Department rolled out a new social networking policy for officers.[75] Under this policy, employees may "not identify themselves directly or indirectly as an employee of the Albuquerque Police Department" or post photographs that contain anything "identifiable to the Albuquerque Police Department."[76] Economidy was returned to patrol duty after over two months at his desk.[77]

Joey Sigala, president of the Albuquerque Police Officers' Association, argued that though it was fine for the department to control what officers did at work, it did not "have the right to tell them what to do outside of that" and that the policy of not disclosing that you were a cop on Facebook also made it hard to share the positives of the job, such as awards or honors.[78]

In New York, a defendant in a felony weapons case, Gary Waters, subpoenaed his arresting officer's Myspace and Facebook posts to support his defense that the cop had planted a gun on him to excuse excessive use of force by the arresting officers. The day before the trial began, Officer Vaughan Ettienne had his mood set as "devious" on his Myspace page. And a few weeks prior to the trial, his Facebook status read, "Vaughan is watching 'Training Day' to brush up on proper police procedure."[79] Officer Ettienne had also made comments on video clips of arrests, including one in which a cop had punched a cuffed suspect. Ettienne commented that if the officer had "wanted to tune [the suspect] up some, he should have delayed cuffing him."[80] He also suggested that the officer had not used that much force and "[i]f you were going to hit a cuffed suspect, at least get your money's worth 'cause now he's going to get disciplined for [it]."[81]

The jurors considering Ettienne's arrest of Waters were swayed by the social network posts. They convicted Waters only of resisting arrest, a misdemeanor, and not of possession of the nine-millimeter Beretta and bag of ammunition found on him, even though Waters was on parole for a prior burglary conviction at the time of the arrest.

Analyzing a cop's subsequent actions based on his posts can be misleading because the posts may bear no resemblance to his actual behavior. Posts give only a narrow view of someone, so it seems unfair to rush to judgment based on posts when other cops may be engaging in actual unethical behavior off-duty or on-duty without any oversight. In addition, people choose to project a certain image of themselves on social networks that may not accurately reflect who they are. (Think about the profiles people post on dating sites, for example.) Can we be sure that if a cop lists *Dirty Harry* as his favorite movie, he will actually use similar tactics on the street? Ettienne told *The New York Times*, "What you say on the Internet is all bravado talk, like what you say in the locker room. You have your Internet persona, and you have what you actually do on the street." Given the potential unreliability of social network posts and the free speech rights of government employees, courts should exclude general tough-guy postings from the courtroom when they are unrelated to a particular case.

Some schools and employers try to create an alternative way of silencing students or workers through contracts such as employee loyalty oaths or student handbooks. When school officials take disciplinary actions against students based on their Facebook pages, they often point to provisions in the student handbook, which sometimes cover behavior out of school as well as in school. Yet do such codes of behavior comply with our society's notions of fundamental freedoms? Consider a Pennsylvania high school's student handbook, which prohibited "inappropriate, harassing, offensive or abusive" speech.[82] When a student posted derogatory remarks about the volleyball team his school would be playing that weekend on an online message board, he was kicked off the team, prohibited from attending any after-school events, and banned from using school computers.[83]

Yet his posts seem like just the sort of things students have been saying for generations about competing schools: "Someone better call the Guiness book of world records, for the biggest lashing in mens volleyball history. These purple panzies are in for the suprise of their lives."[84] In the lawsuit brought by the boy and his parents challenging the school's policy, the court declared that the handbook provision was unconstitutionally overbroad since it could be used to punish legitimate speech that was unlikely to disrupt the school and because it covered social network postings and other activities that occurred outside of school.[85]

Some companies require their employees to agree to act with decorum and not use profanity. This might make sense when the worker is on the job, interacting with customers or other employees, but does it make sense when applied to social network postings?

Just as school codes have been held to be unconstitutionally broad, work codes might also go too far. The legal cases that gave employers the right to forbid profanity arose in the context of factories, where a ruckus among the rank and file could lead to the employment equivalent of a mutiny. As attorney Brandon Brooks points out, profanity about a boss on a social network does not create the same imminent risks.

And duties of loyalty shouldn't override federal employee protection laws, as they did in a case where an employee criticized his employer and was later fired. In that instance, IBM sold a computer circuit board manufacturing facility to Endicott Interconnect Technologies and then became the facility's biggest customer. When, two weeks later, Endicott laid off 200 workers, a local newspaper interviewed Richard White, a longtime employee of the plant and union member. He indicated that the layoffs had left "gaping holes" in the technical knowledge base of the plant. His employer warned him that his comments to the newspaper violated the company handbook policy against disparagement and that if he did it again, he'd be let go. Less than two weeks later, White commented on an anti-union posting on the newspaper's online public forum. He said the "business is being tanked by a group of people that have no good ability to manage it."[86]

When he was fired, the National Labor Relations Board decided that White's right to criticize fell within the protection for speech related to concerted action and ordered reinstatement. But when the company went to court to challenge the ruling, the court mistakenly privileged loyalty over speech. The court held that White's comments were "unquestionably detrimentally disloyal" and "constituted 'a sharp, public, disparaging attack upon the quality of the company's product and its business policies' at a 'critical time' for the company."[87] The court held the termination to be legal.[88] But loyalty shouldn't overrule a person's right to voice concerns about an employer.

Some people may think that workers who post negative comments about their employer are just being stupid, but in a democracy critical speech is valued. Often the speech highlights a safety problem that the public or the investors deserve to know about. And the speech—posts on social networks—usually takes place on the employees' own time and outside the workplace. Those posts are the equivalent, as some union reps have noted, to private conversations around the water cooler or at a bar.

That's not to say that some employee actions aren't beyond the pale. Michael Setzer and Kristy Hammonds, workers at the Domino's Pizza in Conover, North Carolina, filmed disgusting pranks in the kitchen's restaurant and posted them on YouTube.

Unlike the fireman who created an animation related to his job on his own time, the Domino's employees shot their video while on the job. Their cinematic masterpiece showed 32-year-old Michael "wiping his bare butt with a sponge he then uses to wipe a pizza pan, farting on pepperoni, and stuffing cheese up his nose—which is then added to a sandwich under construction."[89] Kristy filmed and narrated the video: "In about five minutes it'll be sent out on delivery where somebody will be eating these, yes, eating them, and little did they know that cheese was in his nose and that there was some lethal gas that ended up on their salami. Now that's how we roll at Domino's."[90]

The restaurant was forced to close due to the videos, and the overall brand was damaged.[91] Within days, the video had scored over a million hits, and the first five of 12 results in a Google search for "Dominos" referenced the video.[92] In that time, the perception of Domino's quality among consumers went from positive to negative, according to the research firm YouGov, which holds online surveys of about a thousand consumers every day regarding hundreds of brands. "We got blindsided by two idiots with a video camera and an awful idea," said a Domino's spokesman, Tim McIntyre.[93]

The filmmakers were fired and indicted on felony charges of distributing prohibited foods, even though they claim the food they shot was never served.[94] Like the nurses who violated confidentiality by discussing their patients' health, the employees were on notice that what they were doing violated an independent law— the state health code. To top it off, their prank occurred on the job, making the firing seem even more appropriate.

The Domino's case seems like an appropriate firing, but how far should we take the principle of not harming the reputation of an employer? What if the kitchen had rats and the two employees had filmed that? What if they made a fictional film, on their own time, where two Domino's workers undertook the disgusting actions? What if they portrayed pizza workers in a fictional film and the employer was not identified or was given a fictional name? Some of these variations deserve to be protected as free speech, as did the fireman's animation. If we allow employers to fire workers for speech that "disparages" their company, we run into the same problems we did with the school codes that ban "inappropriate" or "offensive" speech. Shouldn't a sports fan be allowed to question a team's trade, even if he's employed by the team or the stadium?

Students and employees have been joking and complaining about school and work since time immemorial, mostly to their friends. But when they take their remarks to the internet and social networks, they are creating a digital trail that can lead to unfair expulsions and firings.

Employees and students deserve free speech rights. In some instances, their disclosures provide early warnings of deficiencies in schools or illegal activities at work. The U.S. Supreme Court, in *Pickering v. Board of Education*, for example, held that the First Amendment protected a high school teacher who had been dismissed for sending a letter to a local newspaper criticizing the school board's allocation of school funds.[95] Engaging in concerted activity to change working conditions is a federally protected right, and it would be absurd if, just as social networks come along to make concerted activity easier, employees were punished every time they posted a complaint about their jobs or used social networks to plan strategies to gain workplace rights.

Our Social Network Constitution should protect freedom of expression. At the very least, we should have a rule, such as that carved out under the First Amendment, that protects speech except in cases of clear, imminent harm, such as shouting "Fire!" in a crowded theater (or farting on someone's dinner). But even this approach would seem to give schools and employers the right to snoop around people's web pages in order to see if they are violating the rule.

Such snooping could lead to other harms. If a woman shares an announcement of her pregnancy with her Facebook friends, her boss might use that information as a reason not to promote her, but give a trumped-up reason other than her pregnancy. That action would violate the federal Pregnancy Discrimination Act, but the woman would never know that her boss had learned of her pregnancy and was basing his decision on her reproductive choice. And, in the school setting, if a student says on his Facebook page that he thinks a certain class is boring, the teacher might be tempted to give him a lower grade.

Consequently, our Social Network Constitution should provide even stronger protection for freedom of expression. It should say that social networks are private spaces and that employers, schools and other institutions are prohibited from accessing social network pages or taking adverse actions against a person based on anything they post on a social network. A similar rule is being considered in Germany, which would forbid employers from using information from social networks in the processes of hiring and or judging employees.[96] Finland already bans employers from Googling applicants. The case that triggered the adoption of this rule involved an employer performing an internet search about an applicant and discovering that he had participated in a mental health conference. The employer refused to hire him. Rejecting someone because of a potential mental health problem is bad enough, but the applicant didn't have such a problem—he had attended the conference as a patient's representative. The employer had jumped to an incorrect conclusion based on data from the Web.[97]

Even if employers and schools are prevented from accessing the social network postings of their workers and students, there is still some leeway for punishment in egregious cases, when postings come to the attention of the institution in some other way and indicate the breach of another legal duty. In schools, if a posting incited immediate harm to, say, a teacher or a student, that could be punishable. For employees, a posting could be grounds for action by the employer if the posting itself violated, for example, the duty of confidentiality.

Democracy is based on free speech. The postings of students and employees can provide important information about public and private institutions. But even if their posts do not serve as a form of whistle-blowing, muzzling students and employees would silence people for virtually their whole life span, since people enter school at age five and retire six decades later. Unless the speech is likely to cause imminent societal harm or defames or harasses a private individual, people should be able to freely express their ideas. Such expressions encourage societal discussion. By articulating likes and dislikes, individuals can figure out their own priorities (such as whether they should quit their job) and work to change the conditions of crucial social institutions.

7

Lethal Advocacy

Nadia Kajouji was an 18-year-old student in her first year at Carleton University in Ontario, Canada.[1] She'd graduated at the top of her class in high school and dreamed of becoming a lawyer.[2] An attractive woman with shoulder-length black hair, hazel eyes, and a dark Mediterranean complexion, she arrived at Carleton bright, happy, and ambitious.[3] Her dorm mate Krystal Leonov said, "Nadia, to me, was a very happy person. She was very in depth. Very smart. I always knew that she would do something great in her life."[4]

But Nadia's first year of college was marked by challenges. She fell in love.[5] The condom broke when she and her boyfriend had sex, the morning-after pill failed, and she learned she was pregnant.[6] Instead of talking to family or friends about her problems, she sat in front of the webcam of her computer, sometimes with a blindfold over her eyes, and poured her heart out in a video diary. While recording her words and image, she weighed the tough choice of whether to keep the child or not. She was deeply in love with her boyfriend and could imagine a life with him.[7] But then her boyfriend broke up with her, leaving her to face the tough choice on her own.[8]

Then she miscarried.[9] Now she had no one, not even the baby.

In a restaurant near the university, she pulled out a razor blade and threatened to hurt herself.[10] Someone called the police, who took her to the hospital. But since she was an adult, no one told her parents about the incident.[11]

During winter break, Nadia returned home to Brampton, Ontario, a six-hour drive from school. Her mother, Deborah Chevalier, noticed that Nadia was not at all herself. "Is something wrong?" she asked her daughter.

"I'm just tired."[12]

Back at school, Nadia taped video blogs of herself playing guitar and singing songs of unrequited love. She described her downward spiral to the camera. "I am depressed. I have post-partum mood disorder, clinical depression and insomnia. . . . Maybe I wouldn't be thinking about it if I could sleep."[13]

She told a doctor and a psychologist at the university about her depression and suicidal thoughts, and they prescribed antidepressants, but she still couldn't sleep.[14] Her friends down the hall in the dorm knocked on her door and then tried to contact her through Facebook and through email, but she didn't respond.[15] When Nadia started roaming the residence hall threatening to harm herself, friends called campus security officers.[16] They told the officers that she might be suicidal, but the campus police apparently did not pursue the matter.[17] Campus authorities also never contacted Nadia's parents, although Canadian law permits release of medical information to family if it would reduce the risk of serious bodily harm.[18]

Shutting herself off from her family and friends—perhaps because she didn't want them to see her in such a depressed state, she turned to someone else for comfort. On a social network, she met a young American nurse, Cami, whose email address was falcon_girl_507@hotmail.com.[19] Like Nadia, Cami was depressed. She confessed to Nadia that she'd tried everything from drugs to therapy, from yoga to prayers, and nothing had helped.[20] And now she was thinking of taking her own life. In an email, Cami explained, "about 8 months ago I started looking for methods to let go with and since ive seen every method used possible at work as a emergency ward nurse I know what does and don't work so that is why I chose hanging to use ive tried it in practice to see if it hurt and how fast it worked and it was not a bad experience."[21]

Cami seemed so sympathetic, saying she understood what Nadia was going through and calling her "hun."[22] But Cami was far from a friend. She didn't tell Nadia to seek counseling, talk to her friends or her family, or tell her doctor that her medication didn't seem to be working.[23] In fact, Cami made it seem like none of those would work and that the only remedy was, in the slang for committing suicide, "to catch the bus."

Nadia and Cami chatted back and forth, in hundreds of instant message exchanges.[24] In Cami, Nadia had found someone who understood her and felt her pain.

Or so she thought. In reality, Cami was not the screen name of a young female nurse. Cami was a 46-year-old man, William Francis Melchert-Dinkel,[25] who got his sick kicks out of attempting to convince young women to slash their wrists or hang themselves in front of a webcam so he could watch.

Cami's true identity had been uncovered by Celia Blay. Celia, a retired schoolteacher in Great Britain, had turned to the Web after the death of her parents in search of others who could understand her grief. She stumbled across some of the suicide chat rooms. When she read Cami's posts, she began to suspect that the nurse who was encouraging young people to kill themselves was not who she seemed.[26] From her home in the English countryside, she followed digital clues and discovered the man behind the posts—William Melchert-Dinkel.[27] Pretending to be a young woman, he'd claim to enter a suicide pact with the other person and suggest they both hang themselves in front of the webcam at the same time.[28] Of course, he never held up his end of the bargain.[29] But some of the people he emailed did. Three years before latching on to Nadia, he'd convinced Mark Drybrough of Coventry, England, to hang himself.[30]

But when Celia presented her evidence to the British police and the FBI, they didn't do anything about it.[31] She told them that Melchert-Dinkel had pressured dozens of people to commit suicide and she suspected at least four or five people had.[32] She claims that the local British cops actually told her, "If it bothers you, look the other way."[33] Since Celia knew nothing about Cami's private chats with Nadia, Celia couldn't warn the young woman.

Meanwhile, Cami was pushing Nadia towards suicide by hanging, even telling her what type of yellow nylon rope to purchase at Home Depot in order to get the right size for her body type. But Nadia was going to try to make it look like an accident. She said she would jump into a frozen river with her ice skates on and hope that the rough current would pull her under the ice, drowning her or killing her through hypothermia.[34]

Nadia had stopped going to class by then and recorded another segment of her video diary. "I feel bad wasting my parents' money on school. But I mean, what could I be doing instead? Could I work right now? I don't know."[35] She stopped answering her parents' phone calls.[36] But she continued emailing Cami.[37]

Nadia: So when are you going to catch the bus?
Cami: I would like to soon u?
Nadia: I am planning to attempt this Sunday.
Cami: wow ok you want to use hanging too? Or can u?
Nadia: I'm going to jump.[38]

Cami told her she was going to keep up her end of the suicide pact by killing herself the day after Nadia. And that if Nadia failed in her attempt, they could both hang themselves in front of their webcams.[39]

On March 10, 2008,[40] in the midst of one of the biggest snowstorms ever in Ottawa, Nadia left her room after the other women in the dorm were asleep.[41] She emailed a dorm mate to say she was going ice skating.[42] Leaving her music blaring, she took only her skates, phone, journal, and university access card.[43] In an interchange with Cami before she left the room, Cami did nothing to stop her:[44]

Cami: so you think youll be all done tonite?
Nadia: yup for sure
Cami: ok

Nadia was reported missing the next day.[45] Two weeks after Nadia disappeared, on March 25, 2008, Celia Blay contacted Sergeant William Haider, a police officer with the Saint Paul, Minnesota, Police Department, assigned to the Minnesota Internet Crimes Against Children Task Force. She begged him to look into Melchert-Dinkel, describing him as an "online predator using deception to manipulate parties to commit suicide by hanging."[46]

When the police finally investigated, they learned that Melchert-Dinkel was a nurse who lived on a quiet suburban street in Faribault, Minnesota, with his wife and two teenage daughters.[47] He initially denied that he'd sent the messages.[48] But then, with the police present, he told his wife "Oh I just got into a lot of discussions talking and talking and thinking that I was, being an advocate or helper, or or, God or something or another."[49] Melchert-Dinkel admitted to entering into nearly a dozen suicide pacts[50]—and said that at least five of the users he'd chatted with had later disappeared from the suicide chat rooms, including Nadia and Mark.[51]

For six weeks, Nadia's parents waited in terror to learn if she'd been kidnapped or killed. Then, on April 20, 2008, their daughter's body washed up in a spring thaw of the river. Her body, snagged on a rock in Ottawa's Rideau River, was discovered by a boater.[52]

Nadia's parents wanted Melchert-Dinkel to be brought to justice. But what would be the charge? He hadn't physically pushed Nadia off the bridge. She died in Canada when he was over 600 miles away in another country. And in many places, a person couldn't be held liable for aiding and abetting a suicide unless he was in the same place as the victim and had handed her the lethal drugs or other means of killing herself. Many U.S. statutes put it this way: "assisting suicide" is providing the physical means by which another attempts to commit suicide or participating in a physical act by which another attempts to commit suicide.

Nadia's parents were hopeful, though, since Canada had a law that criminalized the act of counseling, aiding, or abetting a suicide—even if it is done through words alone.[53] But Canadian authorities chose not to extradite William Melchert-Dinkel because they did not believe that his online correspondence with Nadia was a significant factor in her death.[54]

Like Canada, Minnesota, where Melchert-Dinkel lived, had a statute that made it a crime to encourage suicide.[55] But when Melchert-Dinkel was charged by the county prosecutor with encouraging Nadia Kajouji and Mark Drybrough to commit suicide, he argued that his actions were shielded by the First Amendment.[56]

Words are powerful. They can move listeners or readers to action, sometimes even to harm themselves or someone else. But generally, our society doesn't punish the speaker or writer. Think about Ozzy Osbourne. Thirty years ago, he recorded the song "Suicide Solution." The song states that "Suicide is the only way out," and contains the barely recognizable lyrics, sung at a faster speed, "Get the gun and try it; Shoot, shoot, shoot."[57]

When 19-year-old John McCollum shot himself in the head with a .22 caliber handgun after spending five hours listening to Ozzy's music,[58] his grieving parents sued Ozzy and the record distributor.[59] The California Appellate Court rejected their claims, emphasizing the importance of protecting artistic and literary expression and noting that art does not lose its First Amendment protection merely because it "may evoke a mood of depression as it figuratively depicts the darker side of human nature." The court reprinted the lyrics and said they failed to "order or command anyone to concrete action at any specific time."[60] Rather, the lyrics operated as a poetic device with the potential to convey a variety of messages, such as the philosophy that suicide is an appropriate alternative to life.[61] Osbourne's lyrics did not contain the requisite "call to action," the judge said, because no reasonable person would understand them as a command to immediate action.[62]

In another case, Johnny Carson performed a stunt on his show in which he was dropped through a trapdoor with a noose around his neck. Johnny's guest—professional stuntman Dar Robinson—announced, "Believe me, it's not something that you want to go and try. This is a stunt."[63]

But 13-year-old Nicky DeFilippo, Jr., did attempt the stunt and was found hanging from a noose in front of the television, still tuned into the channel that had broadcast the episode. His parents sued NBC, but the judge ruled that the First Amendment barred the DeFilippos from recovering damages. When the family appealed, the state supreme court expressed concern that allowing recovery on the basis of one minor's actions would lead broadcasters to self-censor any material that could be emulated. The court ruled against the parents, suggesting that Nicky was unusually vulnerable, since no one else in the audience copied the stunt.[64]

In the Minnesota case, despite his despicable actions, William Melchert-Dinkel was heading into court with a considerable body of First Amendment law on his side. In 1969, the U.S. Supreme Court had even used the First Amendment to protect the Ku Klux Klan. In *Brandenburg v. Ohio*, the Court overturned the conviction of a KKK leader, Clarence Brandenburg, who'd made a speech to 12 hooded Klan members gathered around a burning wooden cross. Brandenburg had said, "This is an organizers' meeting. . . . We're not a revengent organization, but if our President, our Congress, our Supreme Court, continues to suppress the white, Caucasian race, it's possible that there might have to be some revengeance taken."[65]

Brandenburg was convicted under Ohio's Criminal Syndicalism Statute, which prohibited advocating "the duty, necessity, or propriety of crime, sabotage, violence, or unlawful methods of terrorism as a means of accomplishing industrial or political reform" and voluntarily assembling with any "society, group, or assemblage of persons formed to teach or advocate the doctrines of criminal syndicalism."

When the KKK leader appealed, the U.S. Supreme Court held that the First Amendment protects speech that advocates the use of violence or lawlessness unless the speech "is directed to inciting or producing imminent lawless action and is likely to produce such action."[66] Because the criminal syndicalism statute defined the crime in terms of advocacy, rather than incitement to imminent lawless action, the statute was unconstitutional.

Subsequent cases have held that, for speech to be criminalized, it must be directed to a particular "person or group of persons."[67] The threatened evil must be imminent—speech that advocates action at some indefinite future time is protected by the First Amendment.[68] Under the U.S. Constitution, the government can't penalize speech on the basis that it has a "tendency to lead to violence."[69]

In order to punish Melchert-Dinkel, a judge would have to find that the cyber predator's online chats went beyond advocacy and incited imminent harm. The judge assigned to the Melchert-Dinkel case, Thomas M. Neuville, had spent nearly 20 years in the Minnesota legislature before being appointed to the bench in 2008. The father of five children, Neuville had made crime, family, health care, and cultural issues his legislative priorities. All his areas of expertise came together as he considered the facts of this case.

Judge Neuville rejected Melchert-Dinkel's free speech argument, pointing out that the First Amendment is not absolute. "For example," he said, "the State can prohibit the possession or dissemination of pornographic live performances by minors. Such prohibition is permissibly justified to prevent harm to the minors involved, not to restrict the expressive aspect of the activity."[70]

Similarly, he reasoned, "the government has the right to restrict speech which advises or encourages suicide to foster the government's compelling interest to protect lives of its citizens who are especially vulnerable to suicidal tendencies." He pointed out that Melchert-Dinkel's "encouragement and advice imminently incited the suicide of Nadia Kajouji and was likely to have that effect." The judge labeled the instant messages as "lethal advocacy" and pointed out that Melchert-Dinkel's words were "analogous to the category of unprotected speech known as 'fighting words' and 'imminent incitement of lawlessness.'"[71]

The judge also found that the suicidal predisposition of a victim is not a defense. In the case where the young boy hung himself after watching a stunt on television, the boy's specific vulnerability served as an excuse to let NBC off. But in a case where the inciter targeted a specific person, rather than a general audience, Judge Neuville said that a victim's depression could make her more vulnerable to encouragement and thus make it more likely she'd kill herself.

As Judge Neuville was considering the case, the U.S. Supreme Court decided a case that Melchert-Dinkel's attorney felt might force Neuville to set his client free. In March 2011, in *Snyder v. Phelps, Westboro Baptist Church,* the Supreme Court upheld the First Amendment rights of an anti-gay hate group whose protests at military funerals caused great emotional distress to the family members of the deceased.[72] The protesters' signs said, "Thank God for 9/11," "You're Going to Hell," "Thank God for Dead Soldiers." Even when the soldier whose funeral was disrupted was not gay, the protesters raised signs saying, "Fag Troops" and "Priests Rape Boys." Chief Justice John Roberts wrote the opinion for the Court, saying that the protests, held on public lands, were protected by the First Amendment since they addressed issues of public concern — "the political and moral conduct of the United States and its citizens, the fate of our Nation, homosexuality in the military, and scandals involving the Catholic clergy."[73]

Judge Neuville easily distinguished the suicide case, in which the instant messages were personal, from the general policy discussion of the Westboro protesters. He said, "The speech involved in this case was not public in nature. The speech directed towards the victims in this case was not the subject of general interest or value and concern to the public. It cannot be fairly considered a matter of political, social, or other concern to the community. . . .

"Neither of the victims in this case was a public figure. If Defendant had a strong opinion about the moral, religious, or political topic of suicide or assisted suicide, the First Amendment ensures that he would have almost limitless opportunity to express his view. Defendant may write articles on the topic of suicide. He could create videos and audio recordings and circulate them. He may speak to individuals and groups in public forums and in other private venues that wish to accommodate him. He may appear on television, speak on the radio, and post messages on the internet as long as his speech is of a public nature as determined by all of the circumstances in the case. However, in this case, Defendant focused his advice and encouragement on two vulnerable and depressed victims in a private setting (direct emails and electronic chats). He was not merely advocating a political, moral, or social philosophy."[74]

Under Minnesota law, encouraging suicide carries a maximum penalty of 15 years in jail.[75] But in May 2011, Judge Neuville imposed an unusual sentence.[76] Melchert-Dinkel will serve a 320-day prison term, plus an additional two days on the anniversaries of both victims' deaths each year until 2021.[77] The order requires Melchert-Dinkel to serve a total of 160 hours of community service—8 hours every July (the month when Mark Drybrough committed suicide) and 8 hours every March (the month when Nadia Kajouji committed suicide).[78]

The final blow to the social networks predator: Melchert-Dinkel is not allowed to use the internet without prior court approval.

Was the Nadia Kajouji case rightly decided? Is there such a thing as lethal advocacy? And if so, how should we handle digital aggressors? As we draft our Constitution for social networks, we need to think about how far freedom of expression should go. Do the immediacy and reach of social networks create a different, more deadly sort of speech? And what might we do about it, consistent with First Amendment principles?

Even in the offline world, freedom of speech is not absolute. People can be penalized for spoken and written words, including for assault, bribery, defamation, and fraud. Which way should social network speech cut—should harmful words be thought of as less problematic because thousands of miles might separate the speaker and the victim? Or should they be considered more blameworthy because they can reach people privately in their homes and it's difficult for the victim to determine whether or not the speaker can be trusted? In the funeral protesters' case, the Supreme Court said, "an Internet posting may raise distinct issues" regarding the dividing line between Constitutionally protected speech and harassment.[79] But the Court provided no clue about where to draw the line.[80]

In the Nadia Kajouji case, not only the crime but also the sentence raised First Amendment issues. If our Social Network Constitution contains a right to connect, should it allow a judicial order like the one keeping Melchert-Dinkel off the internet? In some senses, the punishment fits the crime. It seems akin to taking away a driver's license after a DUI vehicular manslaughter. A judge can also condition parole by requiring unannounced visits from a parole officer, and he can prevent the parolee from owning a dangerous weapon.

But judges are limited in the constraints that they can put on a parolee's Constitutional rights. When a judge paroled a female bank robber with the condition that she not get pregnant, an appellate court overturned the condition.[81] Even when women are found guilty of child abuse, a court can't require that they forgo pregnancy since such a mandate interferes with the woman's Constitutionally protected reproductive freedom.[82] For a social network predator, is access to the internet more like a dangerous weapon or more like a fundamental right?

The Nadia Kajouji case caught policy makers' attention since it revealed how speech on social networks can lead to actual harm. From cops to Congress, attention shifted to the prevention of cyberharassment, suicide encouragement, stalking, unwanted sexting, internet defamation, and other means to intimidate and defame people on the Web. Social networks provide a means for people to pry, spy, and lie. But attempts to prevent the harms or prosecute the offenders need to be mindful of the right of free speech.

As law enforcement and public health agencies looked into it, the figures were staggering. A 2010 study involving 10 to 18 year olds found that 20% had been the victims of cyberbullying.[83] The study found that in the previous 30 days alone, 13% of teens had been lied about online, nearly 7% of teens had someone pretend to be them online, 7% of teens had received threats online, and 5% of teens had unflattering photos of them posted online.[84] Adults are victimized as well—targeted by ex-boyfriends, business competitors, political enemies, and even strangers. But existing laws, which assume that only minimal damage can be done by words alone, don't adequately take into account the immediacy, intimacy, and reach of social networks nor the harm that a cyberpredator can unleash.

Attorney Elizabeth Meyer points out how cyberbullying is worse than traditional bullying since it can be done anonymously, it is easier to be cruel from behind a computer screen, the taunting takes place on a world stage, it is more difficult for parents to discover what is going on, and, though children can flee from a schoolyard bully, cyberbullies can reach them anywhere through texts and posts. And when the online posts incite other people to act, the aggressor can distance himself or herself from the harmful actions. Students at a middle school in Southern California responded to a Facebook page entitled "Kick a Ginger Day" (apparently inspired by a similar event on *South Park*) by beating up a young red-haired boy. The boy was beaten by two separate groups of students participating in the digital call to action.[85]

Cruel words delivered on the online playground can have devastating consequences. Megan Meier was a seventh-grade girl living in Missouri who was unhappy in her school.[86] She transferred to a Catholic school and terminated her friendship with a girl from the old school, who lived four houses away from her.[87] That girl's mother, Lori Drew, worried that Megan might be saying bad things about her daughter. So she created a fake Myspace profile, posing as a 16-year-old boy, Josh Evans, and befriended Megan.[88] For weeks, "Josh" flirted with Megan, telling her she was "sexi." Then the tone of Josh's messages changed. One night, he told Megan the world would be better off without her.[89] His Myspace friends started harassing her as well.

Fifteen minutes after Megan received Josh's message, her mother found that she'd hung herself in her closet.[90]

Megan's last message to Josh: "You are the kind of boy a girl would kill herself over."[91]

Missouri prosecutors did not go after Lori Drew because she hadn't broken any Missouri laws.[92] The state's harassment law hadn't kept up with the development of social networks—it punished only harassment via traditional written or telephonic communications.[93] And, way back in 1879, the Missouri law on aiding and abetting suicide had been interpreted by a court to require that the perpetrator be present when the person killed herself.

The inability of Missouri prosecutors to arrest Lori Drew upset many in the law enforcement community. Thomas O'Brien, a federal prosecutor in California, decided to do something about it.[94] He realized that when Drew posed as someone she wasn't, she violated the terms of service of Myspace, which required that people register as themselves. And since Myspace's corporate headquarters was in California, he could claim that the crime was committed there even though Lori Drew and Megan Meier both lived in Missouri.

O'Brien charged Drew with violating a felony portion of the Computer Fraud and Abuse Act that prohibits accessing a computer in excess of authorization in furtherance of a crime or tort. But because the case concerned Drew's use of Myspace, the jury could not consider Drew's activities on other social network sites. The most damning evidence—Drew, as "Josh," telling Megan that the world would be better off without her—had been sent via an AOL instant message rather than posted on Myspace. So the judge asked the jury to deliberate as if that message had never been sent.[95]

The jury convicted Drew of a lesser offense—a misdemeanor—for creating a fictional account using false information in violation of Myspace's terms of service. Drew faced a maximum sentence of three years in jail.[96]

"Trust me, I was so for this woman going away for 20 years," said the jury foreperson, 25-year-old Valentina Kunasz. "However, on the harsher felony charge, it was very hard to find her guilty on the specific [evidence] given to us."[97]

Moments before the court clerk read the jury's verdict, Kunasz glanced towards Tina Meier, Megan's mother.[98] "I honestly wanted to reach out to this woman and give her a hug and tell her I'm sorry—sorry for what she went through and sorry for what she felt and sorry for losing her child," Kunasz said. "The reason I looked at Tina was because I felt like I didn't serve her the justice she was looking for."

Lori Drew was convicted not of cyberbullying, but of violating the Myspace terms of service by not being who she said she was. Despite the fact that she'd gotten off with just a misdemeanor, Lori Drew asked the judge to make a drastic move, rarely used in law—to rule that she was innocent despite the fact that the jury had convicted her.

The tensions in the First Amendment were evident in what happened next. The Electronic Frontier Foundation—a champion of free speech on the inter-net—filed a brief on Drew's behalf. EFF argued that, if Lori Drew could be convicted just because she violated the terms of service of Myspace by not being who she said she was, anybody posting anonymously—even whistle-blowers or people criticizing the government—would become criminals.

Looking closely at Myspace's terms of service, Judge George Wu realized that they included many provisions that ordinary people violated daily. In the judge's mind, if he affirmed the conviction, he'd be making criminal:

1) the lonely-heart who submits intentionally inaccurate data about his or her age, height and/or physical appearance, which contravenes the MSTOS [Myspace terms of service] prohibition against providing "information that you know is false or misleading";

2) the student who posts candid photographs of classmates without their permission, which breaches the MSTOS provision covering "a photograph of another person that you have posted without that person's consent"; and/or

3) the exasperated parent who sends out a group message to neighborhood friends entreating them to purchase his or her daughter's girl scout cookies, which transgresses the MSTOS rule against "advertising to, or solicitation of, any Member to buy or sell any products or services through the Services."[99]

If the computer fraud statute were applied to Myspace violations, any person who breached his or her contract with a social network could be tried as a criminal.[100] That would mean that a major function of government—determining what behavior is criminal—would be put into the hands of social networks.

Judge Wu overturned Drew's conviction and set her free.

Megan's mother was heartbroken. She wanted to see Drew go to jail because she wanted the case to set a precedent: "I think it needed to let people know: You get on the computer, you use it as a weapon to hurt, to harm, to harass people, this is not something that people can just walk away from. This is many times the teen's lifeline."[101]

Because the federal Computer Fraud and Abuse Act proved unhelpful in punishing Drew, some states adopted statutes attempting to define and penalize cyberbullying. But most of these laws have flaws. Some apply only to incidents at school, yet the Centers for Disease Control points out that 65% of young people who reported being cyberbullied were victimized outside of school.[102] Other statutes do not cover the harm that occurs from a single harassing post (rather than repeated harassment) or apply only to bullies who are children. Lori Drew's behavior would not be punishable under any of these approaches since it didn't take place in school, she wasn't a minor, and she pushed Megan Meier to suicide in the single instance of telling her the world would be better off without her. In other states, the cyberbullying laws provide no penalties, but instead require schools to educate students about cyberbullying.

Even in the most disturbing cases, juries seem reluctant to apply the laws. After Lori Drew's conviction was overturned, Missouri broadened its harassment law to penalize anyone who "[k]nowingly frightens, intimidates, or causes emotional distress to another person by anonymously making a telephone call or any electronic communication."[103] When an adult harasses a child on a social network, it is a felony. Otherwise, it's a misdemeanor.

But the new statute provided no solace when it was finally applied to the case of 40-year-old Elizabeth Thrasher whose victim was her ex-husband's new girlfriend's daughter. Elizabeth posted photos, the phone number, and the email address of the 17-year-old girl in the "Casual Encounters" section of Craigslist, in which people expressed their interest in casual sexual encounters.[104] As a result of the "Casual Encounters" posting, the 17-year-old girl, Daniele Pathenos, was swamped with between 20 and 30 sexually explicit cell phone calls, emails, and text messages,[105] which included nude pictures and solicitations for sex.[106] One man even came to the Sonic Restaurant, where she worked, after failing to reach her on her phone, leading her to eventually quit her job out of fear.[107] She testified that the publication of the information made her feel like she "was set up to get killed and raped by somebody."[108]

Elizabeth's attorney dismissed the postings as "tantamount to a practical joke"[109] and argued that photos of the girl and her work location were already available on the girl's Myspace profile.[110] When the jury acquitted Elizabeth, prosecutor Jack Banas argued that Missouri's new cyberharassment statute might need to be stronger yet.[111]

The tactic of using sexual messages to put someone in harm's way is standard on social networks and could be thought of as a new form of sexual harassment. A 2006 study by University of Maryland researchers found that users in a chat room with female user names received 25 times more harassing private messages than users with male names.[112] Rather than being cornered and beat up in a dark alley, women now need to be concerned about being ganged up upon on the Web.

The sexual harassment of women online is reminiscent of the hostile work environments that awaited women in traditionally male professions—such as cops or cabbies or factory workers—in the 1970s. But when a woman's workplace is the Web, cyberattacks can actually run her out of her job. Anonymous posters who want to endanger a woman on the job or off need do no more than post her home address and phone number on Craigslist or a sex site soliciting partners for a rape fantasy. The person putting her in harm's way doesn't even have to leave his apartment. And there is little recourse for the victim of the cyberattack. Unlike in a traditional sexual harassment case, the women and the aggressor are not physically in the same place.

According to the group Working to Halt Online Abuse, 73% of the 349 reports of cyberharassment it received in 2010 involved female victims.[113] Female victims often stop blogging or begin writing under male pseudonyms, foreclosing opportunities for professional advancement like networking.[114] They may also lose revenue generated by advertising.[115] As female blogger Cheryl Lindsey Seelhoff commented, "it is this 'freedom of speech' of some that ultimately silences us and robs us of our own freedom of speech."[116]

Kathy Sierra blogged about how to design user-friendly software, authored a series of books on web design,[117] and created one of the world's largest community websites, JavaRanch.[118] She was a woman working men's turf, commenting on the design of technology.

And that did not go over well with some men. In early March 2007, someone posted "I hope someone slits your throat" on Kathy's blog.[119] Violent and derogatory posts about her began cropping up on other websites, including a doctored photo of her with red lace panties muzzling her face.[120] Someone posted a photograph of Kathy next to a noose. One poster wrote, "The only thing Kathy has to offer me is that noose in her neck size."[121] Another user even posted Kathy's Social Security number and home address.[122]

Afraid to go beyond her own backyard,[123] Kathy contacted local police and canceled an appearance at a technology conference in San Diego, where she was scheduled to deliver the keynote speech.[124] In a statement she posted online, she said that threatening posts should not be covered by freedom of speech: "Are we willing to stake our mother/sister/daughter's life on a sexually and physically threatening photo or comment, simply because it appeared on the internet and therefore must be harmless?"[125]

In early April 2007, Kathy abandoned her Creating Passionate Users blog,[126] which had been listed in the Technorati Top 100 in 2006.[127] Although Kathy stated in March 2007 that she would "lay low no matter what just so I become less of a target,"[128] she eventually resumed some public speaking activities[129] but has not resurrected her blog. She has literally been forced off the Web.

Online harassment can also destroy offline careers. Sometimes the social network predator is not focused on sending disturbing messages to a target, but on circulating derogatory postings about her to cause other people to view her negatively. During the summer of 2005, as she was preparing to move to New Haven, Connecticut, to begin her first year at Yale Law School, Brittan Heller learned that she had become the subject of a message thread titled "Stupid Bitch to Attend Yale Law" on a website called AutoAdmit that hosted discussions by law students, faculty, and lawyers.[130] While the message thread began with "STANFORDtroll" warning Brittan's future peers to "watch out for her," other users began posting sexual threats and false claims about Brittan.[131] "Neoprag" posted, "I'll force myself on her, most definitely" and "I think I will sodomize her. Repeatedly."[132] ":D" wrote "just don't FUCK her, she has herpes."[133]

Brittan had not heard of AutoAdmit[134] before a friend warned her about the disturbing posts. Created in 2004 by Jarret Cohen, a 20-year-old, and co-administered by Anthony Ciolli, a student at the University of Pennsylvania Law School, AutoAdmit's name was a reference to law school applicants who have high enough test scores and grades to be "automatically" admitted to law school.[135] The site attracted between 800,000 and one million visitors each month.[136]

While its focus was purportedly on providing information about law schools and law firms, an astounding number of posts were racist or misogynist. People could post under whatever screen name they chose[137] and AutoAdmit claimed not to keep the ISP addresses of the posters so that it would never have to reveal them.[138] People who innocently came to the site to find out about law schools encountered instead a cyber-cesspool. In 2005, only 150 threads discussed UCLA, just over 100 were about "clerkships," and around 100 were about "Georgetown."[139] In contrast, AutoAdmit contained around 250 threads with the word "nigger," 300 threads with the word "bitches," almost 300 threads with the word "cunt," 350 threads about Jews (the majority derogatory), and over 200 threads about "fags."[140]

The mushrooming posts about Brittan falsely said that she bribed her way into Yale Law School and that she was having a lesbian affair with a Yale Law School administrator.[141] Then the participants in the AutoAdmit network decided that demeaning her wasn't enough. A strand of posts began encouraging others to make sure that the top law firms knew how stupid she was so that she wouldn't get a job offer.[142] Worse yet, since AutoAdmit posts were indexed in Google, the disturbing posts were the first thing potential employers would see about her. Brittan made a list of her top choices for summer employment. All of them turned her down,[143] which a legal journalist pointed out is "almost inconceivable for a Yale Law student."[144]

Brittan was not AutoAdmit's only victim. Heide Iravani, who'd graduated Phi Beta Kappa and with highest honors from the University of North Carolina,[145] became the subject of hundreds of AutoAdmit posts starting in January 2007.[146] One user, "Ugly Women," alleged that Heide had become pregnant after being raped by her dad.[147] "Sleazy Z" urged users to "punch [Heide] in the stomach" while pregnant.[148] Other users expressed their wish that Heide be raped.[149] Another poster, falsely using the name of the law school dean, "Dean_Harold_Koh," as his screen name, claimed that Heide had performed sex on him "for an [sic] P [a passing grade] in Civ pro."[150]

Heide was appalled by the sexual threats and lies posted on the site, including from men who claimed to have forcibly raped her. She worried about how her Muslim father would react. Emotionally distraught, Heide was hospitalized.[151] Then, as with Brittan, the harassers started focusing on her job prospects. In April, an AutoAdmit user sent an email to Heide's upcoming summer employer indicating that there "is some distressing information about her readily available online" that could harm the firm's reputation with clients.[152] By June 2007, the top four Google search results for Heide's name were AutoAdmit postings.[153]

Brittan turned first to Google, asking it to take down the defamatory posts. But

Google's policy is not to remove allegedly defamatory results from its search results, instead directing users to contact the administrators of the websites in question.[154] She contacted the AutoAdmit administrators, but they didn't respond.[155] Heide, too, contacted AutoAdmit, describing her need to seek therapy as a result of the posts.[156] But Anthony Ciolli refused to remove the posting.[157] In fact, he threatened to post any future removal requests he received from her on the site.[158]

So Brittan and Heide exercised some First Amendment rights of their own. They hired ReputationDefender, a business specializing in having defamatory information removed from the Web. ReputationDefender started a campaign on their behalf urging AutoAdmit to respond to complaints about its content and contacted deans at several law schools.[159]

Then, filing as "Doe I" and "Doe II," Brittan and Heide sued 28 pseudonymous posters on AutoAdmit.[160] The poster whose screen name was AK47 argued that compelled disclosure of his identity would violate his First Amendment right to engage in anonymous speech. He and the others who'd posted the threatening and derogatory information had the audacity to argue that unmasking them could hurt their feelings and ruin their job prospects.[161]

The court recognized that although the First Amendment generally protects anonymous speech, including internet speech, the right is not absolute. Anonymous posters can be unmasked in litigation if it looks like the plaintiff can prove defamation,[162] but anonymous posters cannot be unmasked if doing so would stifle legitimate criticism.

Here, Brittan and Heide had been defamed and harmed. The court held that their interest in discovering the identity of "AK47" outweighed his First Amendment right to speak anonymously.[163] So the court unmasked AK47. Some of the other men, such as those who posted as "vincimus" and "A horse walks into a bar" voluntarily disclosed their identities. But most of the men could not be unmasked. They'd posted from internet cafés or other public places or used software to disguise any identifying information.[164]

As the case progressed, the legal arguments made by the attorneys for Brittan and Heide went beyond defamation to include many of the causes of action that grew out of the original 1890 *Harvard Law Review* article, "The Right to Privacy." They included invasion of privacy, revealing information that put the person in a false light, appropriation of name and likeness, and infliction of emotional distress.[165] The case against the pseudonymous posters was settled in favor of Brittan and Heide.[166]

But their emotional saga was not over yet. AutoAdmit is still featuring offensive posts, including a June 2011 racist thread, "I believe I have a great solution to the crime problem in the US," followed by the comment "KILL ALL THE NIGGERS."[167] And when Brittan began working as a postdoctoral fellow with the Afghanistan Legal Education Project in Kabul in 2010,[168] new threads were begun about her, including the posts "Brittan Heller now practicing BIG TERRORIST LAW!" and "We pray for Brittan Heller to have an IED go off in her IUD."[169]

Hostile, threatening posts against women and African Americans create new challenges for the law. Traditional criminal laws might apply when the aggressor causes in-person harm to the victim. Assault charges might be leveled against harassers who threaten people directly. But what about beat-up-a-redhead-day or calls for rape that are sent to other people, where the communication is not with the victim, but with other people or even the whole Web?

Should words alone be punished? It's difficult to determine where free speech ends and criminal activities begin. Is Megan Meier, Lori Drew's victim, more like the boys who committed suicide after listening to Ozzy Osbourne or watching Johnny Carson, totally responsible for her own actions? Or is she more like Nadia Kajouji, who was apparently pushed into harming herself by William Melchert-Dinkel? And how should the law respond to sexual and violent posts against women?

Traditional suicide and harassment laws have loopholes with respect to written threats, and existing criminal and tort law remedies are largely insufficient to address the types of harms caused by online harassment of women and people of color.[170] Posts such as those on AutoAdmit can go beyond harming the targeted individual. Sexual harassment online degrades and alienates females as a group. Law professor Danielle Keats Citron points out that the tort of defamation addresses reputational harms rather than the stigmas that attach to victims of sexual or otherwise demeaning posts. Citron argues for the application of civil rights laws to online harassment of women and minority groups.

In creating a Social Network Constitution with a right to free speech for the Web, we can set some limits. As with the Nadia case, we should be able to hold people liable for imminently inciting harm. Offline bases for lawsuits, such as claims for defamation, privacy invasion, and discrimination, should apply to individuals who post harmful information. Anonymous posters should be unmasked when their postings are likely to cause imminent physical harm, and their identities should be made available when their tirade is defamatory or privacy-invading and about someone in his or her private capacity. Such a limitation would serve to protect anonymous postings about politics, political figures, social institutions, and services (such as restaurant reviews or doctor reviews).

But often the ability to bring a lawsuit against the person who did the posting would be a hollow right. The person might not be traceable, depending on what service or software he or she used to post the comment. And, while the aggrieved person might be able to remove the posting from the original website, it might have already been repeated across the Web. It might also just be replaced by another problematic post on the same website or social network.

Fundamentally, social networks like Facebook and Myspace foster less violent and harmful discourse because most people use their own names. But even in these social networks, some people can be cruel or fraudulent (as in Lori Drew's use of a fake profile to taunt Megan Meier). With social networks and websites that encourage anonymous postings, the potential for harm is even greater and the potential recourse is diminished. In the AutoAdmit case, most of the men who posted violent and defamatory comments could not be identified and thus their deeds went unpunished. The targets of truly anonymous cyberpredators have no remedy.

Can anything be done, consistent with our Social Network Constitution, to try to deter such postings by putting some responsibility on the social networks and websites themselves? Should they be liable when they encourage discrimination, suicide, defamation, or other anti-social acts, as when they feature pages dedicated to those topics or follow a policy of refusing to moderate posts?

The possibility of suing the website itself is limited by a federal law adopted when the internet was in its infancy. Section 230 of the Communications Decency Act shields interactive computer services (including internet service providers and websites) from liability by stating, "No provider or user of an interactive computer service shall be treated as the publisher or speaker of any information provided by another information content provider." In large measure, that act was passed to permit internet service providers to censor certain materials from their sites (such as child pornography) without being viewed as infringing upon free speech. Yet ironically, the act's effect has been to encourage the type of speech that is at the heart of the cyberharassment cases by freeing the websites and social networks themselves from liability for problematic postings.

Privileging internet service providers makes sense in many instances. Comcast, Yahoo!, Gmail, and other services that allow people to connect over the internet couldn't possibly analyze each message sent and censor those that might defame someone or push him or her towards suicide. But are certain social networks—and targeted websites—different? Should some be held to the same standards as publishers or broadcasters—liable for defamation and other torts that anyone publishes in their pages or on their broadcasts?

Celia Blay, the schoolteacher who broke open the Nadia Kajouji case, estimates there are around 7,000 suicide sites on the internet.[171] Nadia and "Cami" met on one called alt.suicide.methods. The social networks and websites devoted to taking one's life provide comparisons of suicide methods in terms of speed, likelihood of success, pain, or likely injuries if the method fails. Some sites contain diagrams of how to aim a pistol into the mouth for maximum effect, recipes for mixing poisons, and resources for obtaining necessary items. A popular Japanese website, describing how to mix toilet bowl cleaner with bath sulfur to create a poisonous gas, allowed users to calculate the quantities of each ingredient for a particular room space and offered a downloadable PDF warning sign to alert emergency workers and neighbors to the poison.[172] Some sites are social network sites where someone with a question can get an immediate answer. When "Overwhelmed in Orlando" wrote, "I'd like to kill myself, but I'm not sure how," the Church of Euthanasia site[173] provided a response addressed to "Overwhelmed" describing exactly how he or she could commit suicide.

Social network information about suicide methods differs from the same information in books[174] because of its ease of access in times of crisis, impression of anonymity, and conduciveness to interactive communications, notably the sharing of personal accounts and the tailoring of methods to the situation of the individual. Plus, joining a social network to find out how to kill oneself may have particular appeal to the population at highest risk of suicide—adolescents.

For the most part, what the suicide-specific social networks do sounds awfully like the protected speech that Judge Neuville talked about in Nadia's case — messages posted on the internet that are directed to everyone, not just one particularly vulnerable person. But when the website provides an immediate answer to a suicidal person or someone takes the conversation with that individual to a private space to encourage suicide (as Melchert-Dinkel did with Nadia), should the website be held accountable for "lethal advocacy"?

And what about a website like AutoAdmit, which encouraged defamatory posts about named individuals? When Brittan and Heide sued one of the administrators of AutoAdmit, Anthony Ciolli, he hid behind the federal statute, Section 230 of the Communications Decency Act.[175] Even though *The New York Times* would be liable for posting defamatory material submitted by someone else, websites that publish defamatory posts are immune from liability under Section 230. As a result, Anthony was dismissed from the case.[176]

Anthony Ciolli then turned around and sued Heide and Brittan for invading *his* privacy and bringing what he said was a frivolous lawsuit against him.[177] That case was settled, with no public disclosure of the terms.[178] Perhaps, though, the women have gotten the last laugh. Despite the many posts aimed at interfering with their job prospects, it was Anthony Ciolli's job offer from a law firm that was rescinded. The managing partner of the firm said Anthony's involvement with a website such as AutoAdmit was "antithetical" to the firm's tenets and the "principles of collegiality and respect that members of the legal profession should observe in their dealings with other lawyers."[179] The partner wrote, "We expect any lawyer affiliated with our firm, when presented with the kind of language exhibited on the message board, to reject it and to disavow any affiliation with it. You, instead, facilitated the expression and publication of such language."[180]

Under our Social Network Constitution, are there situations in which websites themselves should be held liable for defamatory, harassing, or sexually threatening posts that are addressed to a particular individual? This is a tricky question that involves balancing privacy rights against rights of free expression. Immunity from liability makes sense for internet service providers like Comcast or AOL, which merely transmit messages, or general social networks like Facebook or Myspace. But it doesn't make sense with respect to websites whose primary purpose is defamation, invasion of privacy, sexual harassment, hate speech, or incitement of harm.

There is a precedent for stripping social networks or websites of Section 230 immunity when they encourage or participate in the problematic activity instead of merely serving as a mode of communicating the offensive message. The seeds of judicial concern in this area were planted in cases about racial discrimination. Print publications such as *The New York Times* and *Chicago Tribune* are prevented by federal anti-discrimination laws from listing houses under a "No minorities" category. But when a group of Chicago civil rights lawyers sued Craigslist for listing housing that way, the case was thrown out of court.[181] Section 230 immunized Craigslist since it was an interactive computer service.

But does such a result make sense? As communication moves away from newspapers and television to online resources, the need to determine when free speech ends and lethal advocacy begins takes on greater importance. If housing ads have almost entirely moved from newspapers to websites like Craigslist, shouldn't society's decision to thwart racial discrimination move there as well?

In a case decided after the Craigslist discrimination case, federal appellate judge Alex Kozinski, one of America's most notable jurists, found a slight fissure in the Section 230 jurisprudence.[182] The case involved a competitor of Craigslist called Roommates.com. When a person listed housing on that site, he or she could choose from a drop down menu that included "no minorities."[183] For Kozinski, a Romanian refugee who arrived in the United States at age 12 with his Holocaust survivor parents, that menu encouraged users of the site to perform an illegal act, discrimination. He had no trouble finding Roommates.com liable because it wasn't just a passive publisher, but an entity that supplied and directed the offensive content.

But what about the purveyors of other websites that do not have a drop down menu? Under Section 230, no one could sue AOL or Yahoo! if someone posted defamatory or privacy-invading material on one of its Listservs. But when a website exists mainly to incite suicide or defamatory behavior, it should face liability for when victims' lives are cut short or made miserable by the postings. In addition to Judge Kozinski's approach of finding liability when the site facilitates or even encourages the bad behavior, Professor Nancy Kim of California Western School of Law points out that although Section 230 provides that courts can't treat offensive websites and social networks as publishers, it does not say that courts can't treat them as proprietors. She suggests that "proprietorship liability" be imposed on certain websites and that Section 230 immunity be granted only for social networks, websites, and internet service providers that earn it by undertaking certain efforts to prevent harm.

Social networks are, after all, businesses. They sell advertising and sometimes products, such as T-shirts. They provide certain services. If they were bricks-and-mortar businesses, they'd have certain proprietorship responsibilities for what happened on site and off. If a motel fails to install locks on its room doors or does not have lights in its parking lot, that hotel can be found liable if a woman is raped on its grounds. A bar that serves too much alcohol to a patron can be found liable if the patron drives home drunk and runs into someone. And a radio station can be found liable for holding an off-site competition that endangers people.[184]

In her article "Web Site Proprietorship and Online Harassment," Nancy Kim argues that social networks, websites, forums, and chat rooms should be subject to the same reasonableness standard as offline businesses. As part of their duty as proprietors, Kim would require that they take "reasonable measures" to reduce online harassment.[185] Just like the requirement that motels have locks and lights, social networks and websites could be required to do some things to make cyberharassment less likely. Nancy Kim suggests that websites could lessen the online disinhibition effect by measures to deter anonymity, such as by displaying nonanonymous posts higher up on the social network feeds on the website than anonymous ones and by being willing to unmask anonymous posters in certain circumstances. If the poster knows that the barriers for removing anonymity are lowered when he or she posts defamatory or privacy-invading information, he may feel differently about posting harassing content. Kim suggests that easy unmasking of anonymity should be used when "the victim of the online harassment has been identified in a posting; the victim is a nonpublic individual; the victim has signed an affidavit swearing that he or she is the individual identified in the posting; and the victim sets forth facts establishing why easy unmasking is warranted."[186] Because easy unmasking would be limited to cases where the victim's personal information and identity are disclosed online, posters could still discuss matters of legitimate public interest such as general discussion of politics. Proprietorship liability might also require that websites take some measures to moderate content and remove harmful posts.

"CDA 230 has provided vital breathing space for the development and operation of the interactive Internet, but it also leads to perverse and unjust results in particular cases," says Sam Bayard of the Citizen Media Law Project. "It is arguably unjust that a website operator can enjoy the protection of CDA 230 while (1) building a whole business around people saying nasty things about others, and (2) affirmatively choosing *not* to track user information that would make it possible for an injured person to go after the person directly responsible."[187]

Such was the case of the website JuicyCampus, an undergraduate version of AutoAdmit that encouraged students to post defamatory remarks about other students. JuicyCampus encouraged posts about sluttiness, best blow jobs, stupidest Duke student, and claims that certain students were child molesters by touting its immunity: "Please be advised that JuicyCampus is not the author of the posts that appear on the site. Rather, JuicyCampus is the provider of an interactive computer service. As such, pursuant to 47 U.S.C. Section 230, JuicyCampus is immune from liability arising from content posted by users."[188]

Rather than being a passive conduit of information like AOL, the business model of websites like AutoAdmit, JuicyCampus, or exgfpics.com/blog/ (which exists for ex-boyfriends to post nude photos and disturbing critiques of their former girlfriends) is to encourage defamation and invasion of privacy and profit from it.

In New Jersey, prosecutors argued that Section 230 did not shield the Juicy-Campus site from the reach of the state Consumer Protection Act because while the site posted a warning that it didn't allow offensive content, it did not provide any way to get that content removed. They initiated an investigation into JuicyCampus, which shut down its website before the investigation was complete.

Social networks have been a boon to free speech. But they have also created an opportunity for what law professor Brian Leiter calls "cyber-cesspools." The traditional protections against harassment, defamation, invasion of privacy, and discrimination need to be retooled and reinvigorated for the task of bringing justice to the Web.

Although our Social Network Constitution privileges free speech, people need to be prepared to take responsibility for their actions and our society. The very idea of a "social" network seems to be to promote social relationships and interchanges, but the technology itself can foster anti-social actions. Freedom of speech should be limited when it is likely to incite serious harm to another individual. And those limits should apply not only to the offending posters—but to any social networks or websites that act as co-conspirators.

8

Privacy of Place

Every day after school, 15-year-old Blake Robbins followed a common teenage ritual. He entered his bedroom, opened his Apple MacBook, and began surfing the Web. He did a little homework, checked status updates on social networks and instant messaged friends, revealing personal bits of himself with each keystroke. The computer stayed open in his bedroom as his parents and his friends walked into and out of the room, as he returned from the shower and got dressed, even as he slept.

The MacBook was more than Blake's window into the world. It was also the world's window into Blake's bedroom. Unbeknownst to Blake, the camera on his school-issued laptop was peering into his private life.[1] The laptop would take a picture through its webcam, capture a shot of the laptop's screen, and transmit the images to the school district's network every 15 minutes, every time he logged in or out, and whenever the laptop came out of sleep mode.[2]

Five and a half miles away, at Harriton High School in Rosemont, Pennsylvania, members of the Information Services (IS) department reviewed the photos being transmitted by school-issued laptops. Even though the cameras were supposed to be turned on only when a laptop was stolen, the IS department was collecting thousands of photos of students, their family members and friends, and even teachers.

"This is awesome. It's like a little LMSD [Lower Merion School District] soap opera," a Harriton High School IS technician, Amanda Wuest, wrote in an email.

"I know, I love it!" responded Internet Services Coordinator Carol Cafiero.[3]

When the wealthy school district had issued free laptops to its teachers and its 2,300 students,[4] Superintendent Christopher W. McGinley had announced the program with great fanfare. "Every high school student will have their own personal laptop—enabling an authentic mobile 21st Century learning environment," he said. "While other districts are exploring ways to make these kinds of [technology initiatives] possible, our programs are already in place, it is no accident that we arrived ahead of the curve; in Lower Merion, our responsibility is to lead."[5]

McGinley did not tell the students or their parents that the school district had paid an extra $143,975 for LANrev system licenses that allowed laptops to be remotely serviced from the school and included TheftTrack,[6] the webcam program that was activated on Blake's computer. TheftTrack was kept secret until the day Blake was unexpectedly confronted by a school administrator.

On November 10, 2009, Assistant Principal Lindy Matsko called Blake into her office. She thrust a photo at him.

He looked at it in disbelief. *It was a picture of him in his bedroom!*

She pointed at his hand in the photo and accused him of handling illegal pills.[7]

Blake looked more closely at the picture, not knowing what she could be talking about. Then he realized what he was holding. He'd been eating Mike and Ike's candy.[8]

Blake was shocked. "I thought that there was no way that they could do that at my home," he said. His parents were fuming. They obtained the photo and showed it to a reporter.

"Robbins face is plainly visible, in a pose suggesting he is talking to someone else," wrote Michael Smerconish, a *Philadelphia Inquirer* columnist.[9] "He is holding something between his thumb and forefinger that might be a Mike and Ike, it looks the same size and shape. There is no way to be sure. But what's in his hand is irrelevant to the question of whether the school district was justified in creating its own reality show."

Blake's laptop had captured more than 400 images—screenshots of his activity on his laptop while he typed to friends through an instant messaging service, photos taken through the laptop's webcam of Blake while he slept, pictures of him partially dressed after he got out of the shower, and photos of Blake's father and friends.[10] The laptop also transmitted its IP address, allowing the school to track its physical location.[11]

What the columnist found most troubling was the invasion of privacy of unwitting third parties, such as the screen captures of internet chats between Blake and his friends. The texts of the conversations had been redacted, so Smerconish could not tell what Blake and his friends were talking about. "I do know what I was talking about three decades ago before the advent of this technology," Smerconish wrote. "Girls. Sports. Teachers. Administrators. Keggers. And not necessarily in that order."[12]

As word of the Blake Robbins incident spread, other parents in the district—particularly the parents of girls—worried that their children, too, had been spied upon. Savannah Williams, a sophomore at the school, reported that she regularly kept her laptop open in her bedroom while "getting changed, doing my homework, taking a shower, everything."[13] A teacher, Christine Jawork, had already thought there was something creepy about her laptop. After her students had told her that the green light on their laptops started going on for no reason, she had placed tape over the camera on her own school-issued computer.[14]

The Harriton High School community was right to be concerned. The Information Services Department had amassed 30,564 photos through the webcams and 27,428 screenshots.[15]

Blake's parents were furious. Spying on people in their homes has to be some sort of crime, they thought. But once again, the laws were not up to the task of regulating social networks and digital devices. The FBI investigated whether the school district had broken any criminal wiretap laws but declined to file charges. U.S. Attorney Zane Memeger said, "For the government to prosecute a criminal case, it must prove beyond a reasonable doubt that the person charged acted with criminal intent. We have not found evidence that would establish beyond a reasonable doubt that anyone involved had criminal intent."[16]

Senator Arlen Specter was as shocked as the Robbins family that surreptitious videotaping of people in their home was getting a free pass. Lamenting that the Robbins incident "raises a question as to whether the law has kept up with technology,"[17] Specter introduced the Surreptitious Video Surveillance Act of 2010 to amend the federal Wiretap Act to clarify that it is illegal to capture silent visual images inside another person's home.[18] Senator Russ Feingold, a co-sponsor of the act, commented, "Many Americans would be surprised to learn that there is no federal statute to protect them from being secretly videotaped in their homes."[19] The proposed law died without being passed.

With no criminal prosecution possible, the Robbins family took their case to the civil courts. Blake and his parents, Michael and Holly Robbins, filed a lawsuit against the Lower Merion School District on February 11, 2010, seeking damages for the invasion of their privacy and requesting an injunction to stop the school district from spying on any of its students or teachers.[20] They also sought refuge under the laws that people fed up with cookies on their computers had evoked—unlawful interception of electronic communication under the Wiretap Act[21] and unauthorized acquisition of stored electronic communications in violation of the Stored Communication Act.[22] But those statutes had fallen short in the past since courts had held that intercepting or accessing digital communications was not "unlawful" if one party to the electronic snooping had authorized it—and here, the school district had authorized it. So the federal statutes were not likely to help much if the Robbins' case went to trial.

Within hours of the filing of the suit, school personnel began the digital equivalent of shredding evidence. On the orders of the school superintendent, they not only turned off TheftTrack on the tracked computers but they also started deleting photos and other data they'd collected.[23]

The legal action forced the school to notify others that they'd been spied upon and required the school to allow students and their parents to view the remaining images. Jalil Hasan learned that more than 1,000 images had been transmitted to the school by his computer—469 webcam photographs and 543 screenshots over the course of two months.[24] The photos showed him in his bedroom and with friends and family members. The surveillance had not stopped until the day that Blake had filed his lawsuit. "When I saw these pictures, it really freaked me out," Jalil said.[25]

Fatima Hasan, his mother, told *The Philadelphia Inquirer* that she'd moved to this picturesque suburban school district from Philadelphia so her son "would be in a safe environment" for school.

"But then, when I'm looking at these pictures," she said, "and I'm looking at these snapshots, I'm feeling like, 'Where did I send my child?'"[26]

Jalil hired the Robbins family's lawyer, Mark Haltzman, who filed a second lawsuit saying "had the Robbins' class action lawsuit not been filed, arguably Jalil's laptop would have continued whirring away snapping photographs and grabbing screenshots each time it was powered up."[27]

Both the Robbins and Hasan families were outraged that the sanctity of their homes had been violated. And considered in light of basic constitutional principles, their right to privacy was part of the fabric of a free society. Much of the U.S. Constitution's Bill of Rights was dedicated to protecting that right.

The school district leaked to the press that its legal fees had reached $1.2 million to defend the suits.[28] The tactic worked well, as hundreds of families, worried that their taxes might go up, took issue with the Robbins family for filing suit. Rather than expressing concern that their children had been surreptitiously photographed, those parents advocated forgetting the incident once the school turned off the cameras.[29]

But the principle at stake—the unprecedented violation of the privacy of people in their homes—was enough to bring the American Civil Liberties Union into the dispute. The ACLU filed an amicus brief.[30] "The right to privacy inside one's home is sacrosanct,"[31] wrote the ACLU lawyers, quoting federal appellate cases. "The right of a man to retreat into his own home and there be free from unreasonable governmental intrusion stands at the very core of the Fourth Amendment."[32]

The ACLU, which has a thousand members in Pennsylvania, including some whose children attend schools in the Lower Merion School District, focused the court on the long line of technology cases, including the U.S. Supreme Court's decision invalidating the use of heat-sensing devices without a warrant to determine if a person was growing marijuana in his home. "Surreptitious video surveillance inside the home is far more intrusive and revealing than the infrared, thermal detection unit at issue in *Kyllo* [the marijuana case]. And, in fact, the 'extraordinarily serious intrusions into personal privacy' caused by video surveillance has prompted some courts to require the government to justify such searches 'by an extraordinary showing of need.' At least one Court has termed such surveillance 'Orwellian.'"

The school administrators claimed they hadn't known about the ability of the computers to spy on students. But the remote tracking abilities had been described when the administration authorized the purchase of the computers and Theft-Track. And Information Services personnel had put a digital folder of Blake Robbins' photos on the assistant principal's computer, which led to her confronting him about whether he was holding pills.[33]

Some students were aware of the TheftTrack software and had voiced their concerns. In 2009, two members of the student council confronted their high school principal to voice their privacy concerns regarding the software.[34] *The Philadelphia Inquirer* reported that "At a minimum, the student leaders told the principal, the student body should be formally warned about any surveillance."[35] The principal failed to make a schoolwide announcement of the policy or any other change.

Mark Haltzman, the attorney for Blake and Jalil, uncovered evidence that a student intern working in the Information Services Department had also raised the dangers of tracking with the administration before the laptops were distributed. The student intern had emailed the head of the Information Services Department, Virginia DiMedio, and said that it was "startling" that the software would allow the school to remotely manage computers and that parents and students should be warned. DeMedio wrote back:

> What I do know for absolute certainty is that there is absolutely no way that the
> District Tech people are going to monitor students at home. . . . I suggest you take
> a breath and relax.[36]

DiMedio sent the email chain to network technician, Michael Perbix, who wrote to the student:

> The "Big Brother" concern is a valid one. But, I assure you that we in no way
> shape or form employ any Big Brother tactics ESPECIALLY with computers off
> the network.[37]

Because of those false assurances — and a reminder from Perbix that he should not reveal confidential matters that he'd learned in his internship — the student did not raise his concerns to the larger community. Within a few months of those assurances, though, IS staffers were commenting on the "soap opera" goings-on they were able to view in the remote access photos.

Even the school district's own legal team described the Information Services Department as the "Wild West" because "there were few official policies and no manuals of procedures, and personnel were not evaluated regularly."[38] Certain IS employees took the use of TheftTrack lightly and seemed to enjoy the voyeuristic nature of the program.

Not only had students been spied on, so had teachers. One teacher had actually asked the IS folks about disabling the camera, and they wouldn't let her. On 12 laptops issued to teachers, 3,805 photos had been taken, as well as 3,451 screenshots.[39] Some of that material had been collected in connection with a laptop theft, but the system had been activated for no reason at all on half of those teachers' laptops.

Three months after Blake's lawsuit was filed, the school district agreed to a court order permanently stopping the remote activation of webcams on student laptops. The district also promised to destroy all the images captured by the webcams after the students and parents had a chance to view them.[40] Another five months passed before the school district, in October 2010, decided to avoid a trial and paid a settlement of $610,000 to resolve the two lawsuits, most of which went to the lawyer for his work on the case.[41]

But around the world, the capability of video surveillance—coupled with social networks—continued to grow. At Rutgers University that same fall, freshman Tyler Clementi asked his roommate, Dharun Ravi, to give him some time alone in their room one evening. Before he left the room, Dharun turned on his computer's webcam. Then he wandered down the hall to his friend Molly Wei's room. By logging on to Skype on Molly's computer, he and Molly could spy on Tyler and his date.[42]

Dharun felt he'd hit gossip gold. He tweeted to his 150 followers: "I went into molly's room and turned on my webcam. I saw him making out with a dude. Yay."[43]

Two nights later, on September 21, 2010, Dharun tweeted, "anyone with iChat, I dare you to video chat with me between the hours of 9:30 and 12. Yes it's happening again."[44] As Tyler and his lover engaged in this most private of experiences, Dharun and his friend Molly watched from her room.

When 18-year-old Tyler found out, he was devastated. A quiet blond violinist, he reached out on social network sites for advice about what to do. Should he confront the roommate? Alert the administration? Either of those options would bring additional discussion of his sexual orientation.

In the meantime, friends on Dharun's Facebook page were expressing sympathy for his having to room with a gay man. Tyler was shocked when he saw the support that Dharum was getting on Facebook. Tyler expressed his dismay on a gay chat forum: "People have commented on his profile with things like 'how did you manage to go back in there? are you okay?' and the fact that people he was with saw my making out with a guy as the scandal, whereas I mean come on . . . he was SPYING ON ME . . . do they see nothing wrong with this?"[45]

On September 22, 2010, Tyler Clementi drove an hour from campus to the George Washington Bridge. From his car alongside the bridge, Tyler posted a Facebook status update: "Jumping off the gw bridge sorry."[46] Ten minutes later,[47] he leapt to his death.[48]

Police pondered what to do. Had Dharun and Molly committed a crime? Was it cyberbullying? A hate crime? Murder? Even if Senator Specter's Surreptitious Video Surveillance Act had been enacted into law, it wouldn't have protected Tyler because the recording was conducted by his roommate, who had legitimate access to their shared space, entitling him to allow any videotaping there.

As a first step, in September 2010, prosecutors brought criminal charges against Molly Wei and Dharum Ravi for invasion of privacy. Molly Wei was accepted into a pretrial intervention program for first-time offenders and ordered to perform 300 hours of community service, undergo counseling or training in cyberbullying and alternative lifestyles, and work full-time (or part-time if she is in school).[49]

Tyler died in the fall, but it wasn't until the school year was coming to an end that the real legal action began. Molly agreed to testify against Dharun. So, in April 2011, prosecutors filed an indictment against Dharun, charging him with criminal invasion of privacy, intimidation due to sexual orientation, and tampering with evidence (by deleting his tweets inviting people to tune in).[50] Dharun has pled not guilty and is free on a $25,000 bond.[51]

Each day, people Tyler's age from around the world post tributes to him on a Facebook page in his honor. Over 138,000 people have joined this memorial. Although they never met him in life, they comment on his talents, how he has inspired them, and how they are now fighting bullying. And they tell him they hope he's still playing his violin, wherever he is.

While tech insiders like Sun Microsystems' Scott McNealy may say, "You have zero privacy anyway. Get over it,"[52] anyone who is appalled by what happened to Blake Robbins and Tyler Clementi is certainly not willing to get over privacy. Both Blake and Tyler were entitled to a legitimate expectation that their private activities would not be broadcast. But how far does the concept of privacy go in the age of social networks?

The privacy of people in their homes is central to the U.S. Constitution. The Fourth Amendment provides that "The right of the people to be secure in their persons, houses, papers, and effects, against unreasonable searches and seizures, shall not be violated."

But not just homes and dorm rooms are protected. The U.S. Supreme Court has held that the Fourth Amendment "protects people not places." As highlighted by the case where the police had wiretapped the public phone booth Charles Katz used to place illegal bets, the courts have been willing to protect any activities where the person has a reasonable expectation of privacy.

Are social networks public or private? If a woman invited 20 people into her home, the police would not be able to enter without a warrant based on probable cause. If her boss wasn't invited to the party, he would not be permitted to eavesdrop electronically on the conversations there. But what if a woman has 20 Facebook friends? Is that a private gathering or a public one where any third party—from the cops to the boss—can use whatever the woman says against her?

And what should we make of the technologies that permit our whereabouts to be traced or recorded—the cameras on our phones and computers, the smartphone apps that allow the analysis of background noise to determine our locations, the coding of digital photos on our phones and computers that indicate where the photo was taken?

Some people disclose place information, such as their home address, on social networks not realizing the potential consequences. British spymaster Sir John Sawers became a not-so-secret agent in 2009 when his wife, Shelley, posted photos on Facebook showing him in a Speedo, identifying relatives, and indicating where they lived.[53] As the incoming head of the MI6 agency, Sawers's personal information was being highly guarded by the government. But a routine Facebook action put him and his family at risk and led some British politicians to call for his resignation. Ultimately, Prime Minister Gordon Brown allowed the appointment to go through.[54]

In New Hampshire, a burglary ring hit more than 50 homes when people posted status updates on Facebook indicating that they weren't home.[55] Checking in on Facebook and other social networks may lead to robbers checking out your place. There's even a computer program that searches for the word "vacation" on social networks that thieves can use to target homes while their owners are out of town.

In other instances people don't realize what they might be revealing through their social network postings. Photos taken with an iPhone, for example, have embedded in them a string of digital data known as a geotag. When you post a photo of your dog doing tricks, your new engagement ring, or the car you have for sale, the geotag reveals the physical location where the photo was taken. Free software programs can readily decode the information and provide a Google map of the location, leading security analysts to warn about a new problem, "cybercasing," where anything from a theft to the kidnapping of a child can be planned based on data you unwittingly reveal.[56]

Some data aggregators that sell personal information about people also reveal physical locations. Spokeo provides people's home addresses and phone numbers for free and charges a modest fee for additional information, such as access to photos and videos that the person has posted elsewhere on the Web. And in 2011, Facebook started sharing home addresses and cell phone numbers of its users with third-party websites and developers, which flew in the face of users' expectations that the information would be available just to their friends.[57] In a letter to Facebook, U.S. Congressmen Edward Markey and Joe Barton expressed concerns about the move.[58] Facebook halted the sharing, but Marne Levine, Facebook's Vice President of Global Public Policy, has said that Facebook will reenable the sharing of users' addresses and cell phone numbers with third-party applications and websites sometime in the future, after enhancing users' control of the sharing.[59]

Then there are the cyberharassers who maliciously provide your location, putting you in harm's way. That's what Elizabeth Thrasher did when she posted the work location of the 17-year-old daughter of her ex-husband's new girlfriend in the Casual Encounters section of Craigslist.[60] Men started stalking the girl, since the post made it seem that she was interested in casual sex. And when prosecutors went after Elizabeth, she successfully defended her actions by claiming that location information, such as where the girl worked, was available on the girl's Myspace page.

A right to privacy of place in our Social Network Constitution would provide protection whether disclosures are intentional or unintentional and would be in keeping with America's long history of protecting privacy. It would also be consistent with the privacy of place protected by the Fourth Amendment of the U.S. Constitution, which guarantees the security of people's "persons, houses, papers, and effects."

Social networks should be viewed as private places, and thus, under our Social

Network Constitution, third parties would not be able to claim that your location was already public if it appeared on a social network site. Elizabeth Thrasher, Facebook, and Spokeo could be sued or prosecuted for communicating a person's location information without the individual's permission. Additional penalties would apply for actually invading a person's physical space through a webcam or other device transmitting footage from that webcam directly over the internet.

Our Social Network Constitution should also contain due process rights of notice and consent which would require the entities that offer apps or devices to warn you if your use of the app or device will reveal your location. You should be given the chance to easily disable that feature. Currently, disabling geotags requires scrolling through multiple menus and may have the effect of turning off other GPS features, such as Google maps, that the user might still want.

Most of us wouldn't invite complete strangers to our home, or agree to have our sexual encounters broadcast, or publicize the location of expensive items that we own. But social networks, in conjunction with digital devices, can do just that. Only by declaring a right to privacy of place in our Social Network Constitution can we be sure that our desires for safety and privacy and our control over our intimate moments are fully protected.

9

Privacy of Information

On her summer vacation, a 24-year-old high school teacher, Ashley Payne, made a stop at the historic St. James's Gate Brewery in Ireland.[1] Among the photos she posted to her Facebook page was one showing her drinking a glass of Guinness at the brewery. Even though she'd set her Facebook settings to private,[2] someone claiming to be a parent sent an anonymous email to the superintendent of the school district where she worked.[3] The email sender said she was upset about the inappropriate content on Ashley's Facebook page—the Guinness photo and a status update that said Ashley was "going to play Crazy Bitch Bingo" (the actual name of a game played at an Atlanta restaurant, Joe's on Juniper).[4]

When Ashley arrived at school on the morning of August 27, 2009—less than two hours after the superintendent received the email—she was ushered into a meeting with the school principal.[5] The principal told her to resign because of her Facebook posts or he'd suspend her, which could prevent her from ever teaching again. He told her she had to make her decision before she left his office. He told her that she "could not win this." She resigned immediately.[6]

A few months later, Ashley sued the school district for violating the Georgia Fair Dismissal Act because the principal had not told her she was entitled to a due process hearing and that he only had the power to suspend her for a maximum of ten days.[7] She's still fighting that suit. "I did not think that any of this could jeopardize my job because I was just doing what adults do and have drinks on vacation and being responsible about it," she said.[8] The Georgia Professional Standards Commission agreed with her. The commission, which investigates ethics complaints against teachers and can bar them from the classroom, felt that the Facebook photo was no reason to sanction her.[9] But since Ashley had technically quit, it didn't have the power to reinstate her.

Should Ashley's private photos have been cause for public alarm? Did the fact that the pretty young blonde was raising a beer in an otherwise sedate photo mean that she was incompetent to teach? She'd posted over 700 vacation photos, only ten of which showed her with alcohol.[10] Yet the "parent" who had anonymously emailed the school superintendent alleged that her teenage daughter had said, much to the parent's horror, "Mom, tonight I'm going to hang out with my bitches." And that she wouldn't have said that if she hadn't read it on Ms. Payne's Facebook page.

But Ashley's Facebook page was set to private and she had not friended students or their parents. And the statement didn't ring true. Is it really possible that there was an Atlanta teenager who had never heard the term "bitches" except on a teacher's Facebook page?

In fact, there's no evidence that the anonymous email really came from a parent. The account from which it was sent was taken down soon after the email was transmitted. *Atlanta Journal-Constitution* reporter Maureen Downey closely analyzed the language in the entire, lengthy email to the superintendent and speculated that it came from another teacher who didn't like Ashley and wanted her fired. "There are many reasons why I believe that," Downey wrote. "First, very few people outside of teachers have the punctuation skills of this writer. Note the punctuation inside a quote. And I have never had anyone feel compelled to explain that Facebook is 'a social networking site' outside of academics.

"Plus, when is the last time a parent talked about alliteration in such casual fashion?[11]

"As well, most parents would say, 'My child is a student in Ms. Payne's literature class.' But this person wrote, 'My daughter is a **pupil in one of Ms. Payne's literature classes.**'

"Why? Because as a teacher, this person knows that teachers teach multiple classes and falls into that phrasing instinctively."[12] So the anonymous poster was likely not a parent at all but another teacher with a grudge against Ashley.

Ashley Payne is among thousands of people whose innocuous-seeming postings have been used to harm them. Leaks in social network information have led to people divorcing, being fired, being denied admission to college, and committing suicide. Schools, employers, mortgage brokers, credit card companies, and many other social institutions seek information from social networks in order to make judgments about people.

In a 2008 survey, one in ten college admissions officers said they had visited an applicant's social networking site as part of their decision-making process.[13] Some had done so on the invitation of the applicant, while others viewed the site because it was public; 38% of those viewing an applicant's site said it had a negative impact on their admissions evaluation.[14] One respondent had rejected an applicant who had a comment on his social networking profile in which he "bragged that he felt that he had aced the application process for that school, and also that he didn't feel that he wanted to attend that school."

Over a third of employers, according to a CareerBuilder study, say that they will not hire someone whose Facebook page includes photos of that person drinking or in provocative dress. Recently a college grad applied for a job with one of the major talent agencies. His scheduled interview was canceled at the last minute with a phone call saying that no agency would hire him because of a drinking photo on his website. Isn't that amazing? The very type of people who manage hard-drinking film stars—think Charlie Sheen—and undoubtedly consume alcohol themselves—are going to pass on him because of a Facebook photo?

Plus, there's no way for the person to redeem himself. Even if he removed the photo or shut down his page immediately, the photo could still be found on the internet. The Wayback Machine (www.waybackmachine.org) is a project that takes screenshots of websites across time and archives them to preserve a record of the internet as it existed in the past. So that underage drinking photo of you in high school or those rants about your boring job or horrible father might still be available in the archives of the Wayback Machine. A lawyer was fired from a large law firm because he'd posted seemingly misogynistic lyrics on his social network page—six years earlier!

In Bozeman, Montana, starting in 2007, if you applied for a job as a cop or fireman, you were asked to list all your social network accounts—and your passwords![15] Bozeman abandoned the policy on June 19, 2009 in response to the community backlash.[16] But other employers have started to similarly pry into people's second self.[17] Robert Collins, a corrections officer in Maryland, took a four-month leave of absence following his mother's death.[18] Before he was allowed to return to his job, he was required to submit to a Department of Corrections interview. Collins said the interviewer "began to ask me which networks I had the social media accounts on. Then he began to request user name and password information, personal log in information for this stuff."[19] Collins said, "I felt like if I didn't comply completely with the process I wouldn't get my job back."[20] He surrendered his Facebook password and was reinstated.

Collins contacted the American Civil Liberties Union, which forced the Maryland Department of Corrections to investigate the matter. The department indicated that applicants would be informed orally and in writing that disclosures regarding their account were voluntary.[21] The ACLU maintained that in the context of applying for jobs, the department's policy change does little to make this disclosure truly voluntary—it remains coercive.[22]

People post their most intimate thoughts on social networks, acting as if they were in a private conversation with a friend. Most users would be shocked to learn that social networks are considered by social institutions and courts as public spaces, not private ones. Or that privacy-invading assumptions are made about users based on aggregate data from social networks. An MIT study, for example, created "Gaydar," using an algorithm regarding a person's social networks friends to predict, with 78% accuracy, whether that person was gay.[23]

Sometimes people don't realize how much they are unwittingly revealing about themselves through uploads to the internet. Fitbit is a portable device that fitness buffs clip to their clothes to keep track of what they eat and what calorie-burning activities they engage in. The device contains a 3D motion sensor that indicates how many calories the person is burning during each activity. When a Fitbit user passes a Fitbit receiver in her home or health club, her information is uploaded to the internet and added to her written logs from Fitbit's website. According to Fitbit co-founder James Park, "sharing information is an important motivator for [users] to achieve their fitness goals."[24]

Fitbit users had no idea their logs were public and would show up in Google searches. For some 200 Fitbit users, the upload meant that embarrassing details about their lives—from whether they broke their diet to when they had sex, how long it lasted, and how much effort they exerted—could be retrieved under their Fitbit identities by Web search engines.[25] One user logged on June 20, 2011, "Sexual activity. Active, vigorous effort. started at 11:30 pm. 1 hour 30 minutes."[26] But what if that user's spouse was out of town and she was having sex with a lover? Or what if her boss or a family member came across that information? With technologies tracking a person's every move, people may inadvertently share information with the world that they would never have felt comfortable sharing with their closest friend or their trusted psychiatrist.

Virtually every interaction a person has in the offline world can be tainted by social network information. Cops, for example, are finding that their boisterous bravado on their Facebook and Myspace pages can be used by criminal defendants to claim police brutality or police corruption. At a Chicago law firm, a partner ushered the new associates into a conference room on their first day. Around the room, he'd hung up the most embarrassing photo of each of them that he'd found on social networks. "This is your future," he said.

A 54-year-old teacher was on her third round of antibiotics because of yet another respiratory illness she'd caught from her students. She described her frustration on Facebook to what she thought was the limited audience of her friends and joked that the kids were "germ bags." She lost her job as a result.[27] She hadn't realized that the default setting of Facebook had changed to make things public, rather than private.

A flood of evidence is entering the legal system straight from social network pages. Traditionally, courts have allowed discovery about a person's health or behavior when he or she puts it at issue in a case. If I sue you for rear-ending my car, claiming that it gave me whiplash, your attorney can obtain my medical records to see if I had neck problems before the accident. If I am seeking custody of my child in a divorce action, my husband's attorney can seek a court's permission to interview my neighbors to see if I'm a good mother.

But in traditional cases, the other side's attorney can't accompany me and my date to a party, interview people who knew me back in second grade, or find out what my favorite movie or book is. Such information is largely considered irrelevant—and is too costly and time-consuming to obtain. Yet without a moment's thought, judges are granting opposing counsel's requests for everything a person had posted on Facebook, which could include photos from that party, information from old classmates, and clues to the person's likes, dislikes, and activities. What's worse, the information might be irrelevant and prejudicial but still be admitted in court. If your ex-spouse's favorite book is the Bible and yours is *Silence of the Lambs*, guess who might end up getting custody? And if you exaggerate to make yourself appear to be healthier or more fearless than you are, that might come back to haunt you as well. You could lose out on a health insurance claim or a worker's compensation case.

Almost every personal injury case now has a social network connection, in which private health information is used in court. After a Wal-Mart employee claimed that a work-related injury had caused him persistent head and neck pain, a Colorado federal court upheld subpoenas for materials from Facebook, Myspace, and Meetup.com to challenge his claims.[28] When a hospital clerical worker's chair collapsed, she suffered such extensive injuries that she underwent four operations to insert rods in her spine and screws in her neck. The defendant chair company, won a court order for access to her present and past Facebook and Myspace posts.[29] The judge held that if photos showed her smiling or traveling, she could still enjoy life and could not have been injured that badly. But why shouldn't someone with a horrible injury still be able to show a stiff upper lip on the Web?

In New Jersey, teenage girls sued Horizon Blue Cross Blue Shield for coverage of medical expenses related to eating disorders.[30] Horizon argued that the disorders were due to emotional and social pressures, not a medical condition. The judge granted Horizon's request for access to the girls' "writings" related to eating disorders, including content from social networking sites and emails between the girls and family or acquaintances. The case settled when Horizon granted broader coverage for eating disorders. But in other instances, a legal order to expose social network information might cause plaintiffs to drop a case altogether to avoid the stigma and emotional distress of having their most private thoughts made public.

In one case, a disabled wife in Wisconsin was asking for spousal support in a divorce. But her husband convinced the court to admit her recent Match.com profile, which stated that she had an active lifestyle. The court denied her request for support.[31]

In a similar Connecticut case, a judge reached a different result. The Connecticut wife suffered from three ruptured disks, permanent nerve damage in her left leg, and frequent migraines. Trying to start a new life, she listed "biking, rollerblading, kayaking, walks and . . . yoga" on a social network site so that people wouldn't feel sorry for her. The husband tried to use that information to show she was not disabled, but the judge correctly assessed it as "puffery" that the wife used "in order to attract social contacts."[32] The husband was ordered to pay partial maintenance (and she had to give him back some of his prized possessions, including his yearbooks, a small white wicker rocking chair, and a Sears table saw).

The previous sexual encounters of a victim aren't supposed to be admitted in rape cases or sexual harassment cases. But many judges admit evidence of sexuality from Facebook and Myspace accounts. Other courts have denied such requests. When Maria Mackelprang filed suit alleging sexual harassment at work, including coercion to commit sexual acts, her employer asked the judge for access to all emails from Maria's two Myspace accounts. The employer wanted access to her social network accounts to argue that if she had been willing to have sex with anyone else, it would have been no big deal to have sex with her supervisor. A Nevada federal judge denied the request. Not only would such access be an invasion of privacy and potentially embarrassing, said the judge, it reflected the faulty idea that other sexual conduct creates "emotional calluses that lessen the impact of unwelcomed sexual harassment."[33]

Personal information about us is routinely pulled from our social network pages, often without our knowing it. We may have signed on to a social network, thinking we were entering a private space, only to have the network pull the rug out from under us and change its rules once we've gotten hooked. (See "Facebook's Eroding Privacy Policy: A Timeline.")

PRIVACY OF INFORMATION

Kurt Opsahl, "Facebook's Eroding Privacy Policy: A Timeline," Electronic Frontier Foundation, April 28, 2010, www.eff.org/deeplinks/2010/04/facebook-timeline. The Electronic Frontier Foundation is a nonprofit public interest group based in San Francisco, California.

FACEBOOK'S ERODING PRIVACY POLICY: A TIMELINE

Since its incorporation, Facebook has undergone a remarkable transformation. When it started, it was a private space for communication with a group of your choice. Soon, it transformed into a platform where much of your information is public by default. Today, it has become a platform where you have no choice but to make certain information public, and this public information may be shared by Facebook with its partner websites and used to target ads.

To help illustrate Facebook's shift away from privacy, we have highlighted some excerpts from Facebook's privacy policies over the years. Watch closely as your privacy disappears, one small change at a time!

Facebook Privacy Policy circa 2005:

No personal information that you submit to Thefacebook will be available to any user of the Web Site who does not belong to at least one of the groups specified by you in your privacy settings.

Facebook Privacy Policy circa 2006:

We understand you may not want everyone in the world to have the information you share on Facebook; that is why we give you control of your information. Our default privacy settings limit the information displayed in your profile to your school, your specified local area, and other reasonable community limitations that we tell you about.

Facebook Privacy Policy circa 2007:

Profile information you submit to Facebook will be available to users of Facebook who belong to at least one of the networks you allow to access the information through your privacy settings (e.g., school, geography, friends of friends). Your name, school name, and profile picture thumbnail will be available in search results across the Facebook network unless you alter your privacy settings.

Facebook Privacy Policy circa November 2009:

Facebook is designed to make it easy for you to share your information with anyone you want. You decide how much information you feel comfortable sharing on Facebook and you control how it is distributed through your privacy settings. You should review the default privacy settings and change them if necessary to reflect your preferences. You should also consider your settings whenever you share information. . . .

Information set to "everyone" is publicly available information, may be accessed by everyone on the Internet (including people not logged into Facebook), is subject to indexing by third-party search engines, may be associated with you outside of Facebook (such as when you visit other sites on the internet), and may be imported and exported by us and others without privacy limitations. The default privacy setting for certain types of information you post on Facebook is set to "everyone." You can review and change the default settings in your privacy settings.

Facebook Privacy Policy circa December 2009:

Certain categories of information such as your name, profile photo, list of friends and pages you are a fan of, gender, geographic region, and networks you belong to are considered publicly available to everyone, including Facebook-enhanced applications, and therefore do not have privacy settings. You can, however, limit the ability of others to find this information through search using your search privacy settings.

Facebook Privacy Policy, as of April 2010:

When you connect with an application or website it will have access to General Information about you. The term General Information includes your and your friends' names, profile pictures, gender, user IDs, connections, and any content shared using the Everyone privacy setting. . . . The default privacy setting for certain types of information you post on Facebook is set to "everyone." . . . Because it takes two to connect, your privacy settings only control who can see the connection on your profile page. If you are uncomfortable with the connection being publicly available, you should consider removing (or not making) the connection.

> Viewed together, the successive policies tell a clear story. Facebook originally earned its core base of users by offering them simple and powerful controls over their personal information. As Facebook grew larger and became more important, it could have chosen to maintain or improve those controls. Instead, it's slowly but surely helped itself—and its advertising and business partners—to more and more of its users' information, while limiting the users' options to control their own information.

Even if we specifically designate our social network pages as private, social network sites do a particularly bad job of protecting privacy. In 2011, Joanne Kuzma of the University of Worcester analyzed the privacy practices of 60 social networks, including Facebook, Myspace, and LinkedIn. The good news is that 92% of the social networking sites in the study were found to have some type of overall privacy policy. The bad news is that 37% of websites were found to contain third-party cookies and 90% contained web beacons. Such privacy invaders are able to track information about how many times the user accesses a page, his browsing history within a site, and any information he types into forms on the site (including credit card information and Social Security numbers).[34] Another study, conducted by researchers at Cambridge University, found that social networks such as Facebook and Myspace made it difficult for users to determine their privacy policy.[35] The authors suggested that social networks do so due to concerns that people would not sign up if they knew how limited the privacy protections were.[36]

"We wouldn't recommend posting anything there that you wouldn't want marketers, legal authorities, governments (or your mother) to see, especially as Facebook continues to push more and more of users' information public and even into the hands of other companies, leaving the onus on users to figure out its Rubik's Cube–esque privacy controls," says *Wired* technology columnist Eliot Van Buskirk.[37]

Privacy on social networks is even a concern of the savvy high tech experts, some of whom feel violated by the way in which Facebook has played fast and loose with its privacy settings. When it started, Facebook gave users complete control over their privacy settings. Then, noted Ryan Singel of *Wired*, "it reneged on its privacy promises and made much of your profile information public by default."[38] After the change, there was no way to make your name, city, photo, or friends private, and causes, likes, and links were public unless you removed them from your profile entirely. Singel expressed his frustration with the lack of control: "I'd like to have my profile visible only to my friends, not my boss. Cannot. I'd like to support an anti-abortion group without my mother or the world knowing. Cannot."[39]

Even trying to assert control of the information that Facebook does let you keep private is a chore. *The New York Times* pointed out in a complicated diagram that

"To manage your privacy on Facebook, you will need to navigate through 50 settings with more than 170 options."[40]

At least with Facebook, you can ultimately find the site's privacy statement and read through its 45,000-word privacy policies.[41] Data aggregators that are collecting information from your social networks pages, your search requests, and your emails don't provide a ready way for you to know you're being targeted.

Advertisers use data aggregation and targeted ads to enhance a company's bottom line by offering you personalized ads. Emails from your Gmail account undergo text analysis for similar commercial purposes. But the potential for analyzing your proclivities is expanding beyond commercial enterprises. Public health officials and psychologists want to use text analysis to identify people searching for information on infectious diseases or how to commit suicide. In the process, they may be causing privacy breaches that do more harm than good when they analyze search queries or social network posts that the individuals thought were free from scrutiny.

Textual analysis is being proposed to identify and try to prevent suicide risk, bullying, and other problematic behaviors. In 2009, researchers at Victoria University developed an algorithm for identifying potentially suicidal users of social networks.[42] The researchers studied suicide notes to amass a dictionary of words and phrases associated with suicide. The researchers then used an automated crawler to assign scores to individual Myspace users' blogs based on the frequency of those suicide related words and phrases. According to the researchers, the text mining technique could be used to direct suicide prevention advertisements to the most at-risk users or to send online messages asking whether at-risk users needed help.

Using text mining to characterize individuals as "suicidal" is problematic given the particular limitations of text-mining and algorithm techniques. Because the algorithm is designed to capture more false positives than false negatives, users who are not suicidal may rank high on the "at-risk" scale. The text-mining techniques are unresponsive to context, capturing irrelevant material such as narratives, quotes, or quizzes containing supposedly suicidal words ("I could kill myself for forgetting my house key!") or negative use of words ("No matter how sick I get, I would never want to kill myself.").[43]

Also troubling is the potential long-term storage of scores in databases. When, if ever, would an individual escape an algorithm diagnosis of "suicidal"? Would suicide prevention organizations or law enforcement more heavily scrutinize users identified by an algorithm as "at risk"? Would employers be reluctant to hire someone with a troubling score?

Spying on social network users, whether for their own good or to fuel a business like behavioral advertising, violates the person's right to privacy. All of a person's posts and photos on social networks should be considered private. The traditional legal test protects privacy when a person has an expectation of privacy, and that expectation is reasonable. Both those criteria are met with social networks. In addition, protecting social networks postings as private is beneficial to individuals and to society.

People were lured to Facebook with the promise that it would be a private space. Back in 2005, when it was still called "Thefacebook," the privacy policy stated, "No personal information that you submit to Thefacebook will be available to any user of the Web Site who does not belong to at least one of the groups specified by you in your privacy settings."

The entire structure of Facebook revolves around affirmatively choosing someone to be part of your circle of friends. Unlike Twitter and YouTube, which are generally meant to be seen by the world at large, social networks foster the notion that you are just communicating with a group of intimates. No matter what your privacy settings are, you need to be asked whether or not someone can friend you, which gives you the sense that you have control over who enters your private digital space, just as you have control over who enters the physical space of your home. If posting to Facebook were like writing on a bathroom wall, there would be no need for a mechanism where people asked whether they could be your friend or not.

The underlying structure of Facebook—the impression that you are speaking to only your friends—is what provides the impetus for people to provide more intimate, personal, and revealing details than have ever before been shared on the Web. People talk about their emotions, their likes and dislikes, their feelings about the people in their life, their goals for their families, their sex lives, their laudatory and blameworthy acts.

People's gut feeling that what they are posting is private is a belief that it is reasonable for society to protect under the right to privacy. Information about sexuality, health, crucial life choices, and personal beliefs is traditionally protected as private—and those are exactly the sorts of things that people post on Facebook and other social networks. Myspace even has an emoticon function where users can readily indicate their emotional mood. Declaring private the intimate information that people post on social networks is in keeping with the fundamental right to privacy articulated by Louis Brandeis, the co-author of the pivotal 1890 *Harvard Law Review* article that created the foundation for current privacy laws.

"The makers of our Constitution undertook to secure conditions favorable to the pursuit of happiness," said Brandeis when he became a member of the U.S. Supreme Court. "They recognized the significance of man's spiritual nature, of his feelings and of his intellect. They knew that only part of the pain, pleasure and satisfactions of life are to be found in material things. They sought to protect Americans in their beliefs, their thoughts, their emotions and their sensations."[44]

Protecting the privacy of information on Facebook and other social networks is also in keeping with the recognition of privacy for places. People access their social network pages privately, often in the home. Cyberspace is a place, and your Facebook page is your little bit of real estate in that place. In fact, for some people it is a more private space than any other he or she has.

Microsoft researcher and social networking expert danah boyd explains that younger people are willing to post personal information online because they feel it is an even more private space than their physical space. "Kids have always cared about privacy, it's just that their notions of privacy look very different than adult notions," she told *The Guardian* newspaper. "As adults, by and large, we think of the home as a very private space . . . for young people it's not a private space. They have no control over who comes in and out of their room, or who comes in and out of their house. As a result, the online world feels more private because it feels like it has more control."[45]

The consequences of people invading your private online space are as grave as those of people invading your offline space. In a thoughtful opinion, the Philadelphia Bar Association likened a Facebook page to a person's home. In 2009, the association's Professional Guidance Committee concluded that lawyers can't ask other people to "friend" an opponent's witness to gather evidence to impeach the witness. The committee was unwilling to invade the privacy of a witness even if that person's social network privacy setting was wide open and even if the person was willing to accede to all friend requests. "Deception is deception, regardless of the victim's wariness in her interactions on the internet and susceptibility to being deceived," wrote the committee. A lawyer claimed that fake friending was no different than videotaping a plaintiff in public in a personal injury case to show she isn't really injured, a common tactic of defense lawyers. The committee disagreed, considering social networks to be private spaces. Although lawyers might be able to videotape a witness in public, they said that friending to get information would be like posing as a utility worker and using a hidden camera to film someone inside her home.

Privacy is not the same as secrecy. As a 2001 court decision recognized, "the claim of a right of privacy is not 'so much one of total secrecy as it is of the right to *define* one's circle of intimacy to choose who shall see beneath the quotidian mask.' 'Information disclosed to a few people may remain private.'"[46]

The judge who held that Cynthia Moreno's Myspace page shouldn't be considered to be private got it wrong. So did the judge who determined that once your email was opened by its recipient, the information it contained was no longer private. The rule that once you tell one other person a private fact, it's the same as publishing it to the world just isn't good law. We protect information from further disclosure in lots of instances in which a person makes an initial disclosure. If you tell your doctor that you have the symptoms of a sexually transmitted disease, courts can't just say, well, the cat's out of the bag, it doesn't matter if *The National Enquirer* or Perez Hilton republishes that fact. If someone sends you a letter complaining about his or her spouse, you can't publish it in your book about marriage unless you receive explicit permission from the sender. In 1986, for example, J. D. Salinger successfully prevented the publication of private letters he'd written to other authors by asserting his copyright over the letters.

The right to privacy on social networks comports with other legal precedents that stake out a personal realm, even if it is in a public space. For example, some states have laws that protect off-duty conduct of employees, such as Ashley Payne's visit to the Guinness factory. Who would have thought that lawmakers out to protect the tobacco industry would have seeded the laws that are workers' best hopes in the social networks cases? But when the push to discourage smoking started, pro-tobacco lawmakers introduced laws, still on the books in at least 17 states, that prohibit employers from discriminating against employees' smoking habits outside of the workplace.[47] But lawmakers in at least eight states went further, adopting laws that prohibit discrimination against employees based on the use of lawful products (or, in the state of Minnesota "lawful consumable products").[48] Some lawyers are arguing that the use of social networks is the use of a lawful product and is thus protected.[49] If that's the case, such a law would protect a worker against being fired when she—or someone else—posts a photo of her drinking. Another four states (California, Colorado, New York, and North Dakota) prohibit discrimination against employees based on lawful activities they undertake on their own time,[50] perhaps providing cover for workers whose embarrassing (but legal) exploits are posted on Facebook. These states bar employers from making adverse employment decisions (ranging from hiring to firing) based on legal activities an employee engages in when he or she is not working.[51]

New York's statute is one of the broadest. It says that an employee can't be punished for his or her "legal use of consumable products or legal recreational activities" outside the office, including "sports, games, hobbies, exercise, reading and the viewing of television, movies and similar material."[52] Since time on social networks is also a recreational activity, it would be covered. The explicit mention of reading and movies would also protect a person's Facebook "likes." The intern at a fashion magazine whose favorite movie is *The Devil Wears Prada* or the young cleric whose favorite book is *American Psycho* could not be fired based on those social network postings. However, the protections don't apply if the employee uses the employer's equipment or other property. This means that if the worker connects to Facebook using an employee-issued phone or computer, that worker could be out of luck.[53]

A few other legal precedents recognize that social networks are private spaces. When a federal law was adopted to prevent employers from discriminating against healthy employees who had a potential genetic predisposition to disease that might later affect them, it banned employers from getting genetic information from the employee's medical records or by ordering the employee to undergo genetic tests. But along came social networks, where an employee's disease might be directly revealed in a post about a trip to the doctor or indirectly if he or she joins a group related to that disease. In 2010, the Equal Employment Opportunity Commission specifically forbade employers from discriminating against workers based on genetic information found on their social network.[54]

Protections for privacy serve important individual and social purposes. Even in our exhibitionist culture, people long for privacy. A 2010 study by Pew Research Center found that though younger adults (ages 18 to 29) tend to be more cautious with their online privacy than older users, 65% of all adult social network users have changed their privacy settings to limit what they share with others online.[55] Many others may just not know how or may assume that their Facebook page is already private.

Control over private information is essential to respect, friendship, love, trust, and personal liberty, according to privacy scholar Charles Fried.[56] We achieve intimacy with other people by parceling out information about ourselves bit by bit. Each new revelation demonstrates additional trust. It's healthy to be able to show different aspects of ourselves in different settings. We need to explore or grow without our former digital lives coming back to haunt us.

Disclosure of private posts and emails can be emotionally devastating. In fact, psychiatrists and philosophers disagree completely with the Facebook founders' notion that all information about a person should be "transparent." Philosopher Sissela Bok says, "Control over secrecy provides a safety valve for individuals in the midst of communal life. . . . Psychosis has been described as the breaking down of the delineation between the self and the outside world: the person going mad 'flows out onto the world as through a broken dam.'"

Privacy is also the way we maintain civility and dignity in the larger society. Robert Post, the dean of Yale Law School, says, "'privacy' is simply a label we use to identify one aspect of the many forms of respect by which we maintain a community. It is less important that the purity of the label be maintained, than that the forms of community life of which it is a part be preserved."[57]

An invasion of privacy destroyed the notions of civility and dignity for the Catsouras family. When 18-year-old Nikki Catsouras took her father's black Porsche 911 out for an afternoon drive, she was the image of a typical teenager—an aspiring photographer who had given her father the peace sign in the house just minutes before she went outside and broke her parents' rule against driving the Porsche.[58] But that Halloween day, Nikki became an image of gore, forwarded in emails and posted on websites.[59] Fifteen minutes after leaving her home, Nikki crashed into a concrete tollbooth at nearly 100 miles per hour—she died upon impact.[60] The responding police officers took 50 photographs of the mangled Porsche and Nikki's nearly decapitated head as part of their routine investigation and later transferred the photos to police office computers.[61] A dispatcher at the precinct, Aaron Reich, emailed nine of the gruesome photos to family and friends,[62] later claiming that the emailed images were intended as reminders of the dangers of reckless driving.[63] Although he sent the photos to only a few people,[64] they were forwarded on and, in the words of the California Court of Appeals, "spread across the Internet like a malignant firestorm."[65] Some 2,500 websites,[66] many devoted to pornography and death, featured photos of Nikki's mangled body.[67] A fake Myspace profile of Nikki was even created, featuring a closeup of Nikki's head and the caption "What's left of my brain here: As you can see, there wasn't much."[68]

Nikki's body had been so mangled that the coroner had not permitted her parents to view it.[69] But a few days after Nikki's death, her father received an email, disguised as a work-related communication, that contained a picture of Nikki's dangling head and the words "Woohoo Daddy!" "Hey daddy, I'm still alive."[70] He soon realized that that the disturbing photos were all over the internet. Fearful of encountering the photos again, he left his job as a real estate agent for one that paid less but allowed him to avoid the internet.[71] He and Nikki's mom forbid her sisters from using the internet, and when students threatened to put Nikki's photos on one sister's locker, they decided to homeschool her.[72]

The Catsouras family attempted to remove Nikki's gruesome death images from the Web. They sent emails to Google, Myspace, and various websites, asking them to take down the photos.[73] A relative spent 13 hours a day for a month pleading with site administrators.[74] But many sites refused to remove the photos, referring to their First Amendment rights.[75] As the owner of one site responded, "When we look upon photographs, like those of a young girl who has been violently struck down in the prime of her life by a moment's recklessness, we gaze upon our own mortality, and we think about how easily this could have been us . . . the right of the rest of us to view such images should not be infringed upon."[76]

Unable to escape the images, Nikki's parents and three sisters filed suit against the California Highway Patrol (CHP), Reich, and another dispatcher, Thomas O'Donnell, who had sent the pictures to his personal email address.[77] The suit alleged that the dispatchers and their department had not met their responsibilities, that they'd invaded Nikki's privacy and that of her family, and that they'd intentionally caused the family emotional distress.[78] The trial court found that, although leaking the photos was "utterly reprehensible," there was no basis for the lawsuit because any rights died with Nikki.[79]

Taking a second mortgage out on their home to pay for the additional legal expenses, the Catsouras family appealed the decision.[80] The California Court of Appeals chastised the CHP and its officers for "deliberately making a mutilated corpse the subject of lurid gossip."[81] Unlike the trial court, the California Court of Appeals found that the CHP officers owed the Catsouras family a duty of care not to place Nikki's photos on the internet "for the lurid titillation of persons unrelated to official CHP business." According to the court, it was foreseeable "that the gruesome photographs allegedly disseminated for shock value on Halloween would be forwarded to thousands of Internet users, in these days of Internet sensationalism." Moreover, the court found that the online release of Nikki's photos intruded upon the Catsouras family's privacy because it constituted a "morbid and sensational prying into private lives for its own sake," rather than for a legitimate public interest or law enforcement purpose. The court also allowed her parents to go forward with their emotional distress claim.[82] Although the Catsouras family's legal battle is not over (the case must now be tried at the trial court level), the California Court of Appeals' decision has opened the door for legal action in other cases of invasion of privacy involving the Web.

Our Social Network Constitution should provide a right to privacy that covers what you and others post about you on a social network, as well as inferences that can be made about you based on what you write, view, and visit on the Web. If we took seriously the right to privacy on social networks, several things would follow. You could control what was collected about you, preventing data aggregators from trolling through your online posts unless you specifically agree to it. And you'd have control over what uses were made of your information. If your social network postings were viewed more like something whispered in the home than like something shouted in the public square, you'd be protected against their being used in discriminatory ways against you. If your employer or school or a credit card company shouldn't be seeing this information in the first place, they wouldn't have a right to treat you negatively based on that information. Privatizing the social network space—irrespective of how open you have made your settings—will help ensure that the individual and social benefits of privacy are realized by preventing your personal information from being used against you.

The circumstances under which that privacy can be invaded, even by the legal system, should be severely limited. Right now, some lawyers and judges consider everything a person has ever posted as fair game in civil cases, custody cases, and criminal cases. The result is that a lot of potentially prejudicial information is admitted into court that can unfairly taint the judgment. A woman who posted something sexy on her boyfriend's Myspace page lost custody of her child to her ex-husband. A kid wearing gang colors on his Facebook page had his criminal sentence enhanced.

Our Social Network Constitution should allow access to evidence on social networks only in extremely limited circumstances and only with a court order. It seems appropriate to allow access if the posting is direct evidence of the crime or other harm at issue, such as extortion, jury tampering, or fraud. But if a person has merely put his or her condition at issue by alleging an injury or disability, that shouldn't provide the basis for the other side to snoop on social networks. Instead, the other side in the litigation should be limited to obtaining evidence of the person's condition through means used in the past—such as medical exams—rather than access to social network posts, which may reveal the most intimate aspects of the person's life.

Recognizing a right to privacy in our Social Network Constitution will bring us in line with other industrialized countries that protect privacy on social networks. Throughout Europe, laws and treaties about data aggregation outlaw the types of privacy invasions we are seeing in the United States. Under the Social Network Constitution, the assumption, the default setting on social networks, should be that no one would have access to your information unless you specifically agree after adequate warnings about the implications of giving up your privacy in a particular online setting and adequate disclosure about who will be using that information and for what purposes.

10

FYI or TMI?: Social Networks and the Right to a Relationship with Your Children

When Geraldine Black filed to obtain custody of her great-grandchild, she had plenty of ammunition. Her grandson, the child's father, was in jail. And according to Geraldine and members of her family, the child's mother was no saint either. After all, the mother's Facebook page contained a post that she'd celebrated "National Weed Day," with the exclamation "Fire it up."[1] The family court judge readily admitted that evidence and took the child from her mom. When the mom appealed, an appellate court gave the child back to her—not because it was wrong to introduce Facebook evidence but because the child's physical health and emotional development seemed fine.

Anything you post can and will be used against you—and that is especially true in custody proceedings. But how well do social network posts reflect an ability to parent? Are all those 616,000 fans of Jimmy Buffett on Facebook necessarily bad parents because he sings about drunken debauchery? How about if they share his song "Why Don't We Get Drunk?" with the lyrics "I just bought some Colombian herb and we'll smoke it all, me and you"? What if a person's Facebook page indicates he likes books by Samuel Coleridge, a known opium addict? Or what if she joins a Myspace group in favor of medical marijuana? Should a man lose custody of a daughter if he watches rap videos on YouTube that have misogynistic lyrics? Or does a Google search for child pornography?

Social networks are transforming how relationships begin and end. One in five relationships now starts on social networks. And social network information can be a smoking gun when people divorce. An American Academy of Matrimonial Lawyers poll found that 81% of divorce attorneys have encountered an increase in the use of social networking evidence over the past five years.[2] Most of that evidence was found on Facebook (66%) or Myspace (15%).

Evidence that one spouse cheated on another or has dangerous habits can help the other spouse receive more money in the split or gain sole custody of the kids.

Divorce lawyers scour social networks to make their case for them. Linda Lea M. Viken, the president of the American Academy of Matrimonial Lawyers, recounted a situation during a custody battle where a father posted on his Facebook page that he was "single with no children looking for a fun time."[3] Information such as this is important to matrimonial lawyers because, as divorce lawyer Kenneth Altshuler said, "Facebook has made it very easy to show lack of credibility and that is what can win a case. Once you catch them in one lie, nothing else they say is credible to the judge."[4]

Are people in Facebook Nation more likely to cheat than before the advent of social networks? It's easier to meet people online than offline, and people can now connect with former high school boyfriends or other ex-loves to rekindle old romances. Websites like ashleymadison.com facilitate adultery by exclusively targeting married people who want some action on the side.

But as with other aspects of social networks, it's unclear whether there's more bad behavior—or just more evidence of it—than in previous generations. Instead of catching a whiff of another woman's perfume on your husband's shirt, you might instead find an X-rated photo that your husband accidentally tweeted to a woman in public mode rather than private mode. Or—as happened in a Connecticut case—your husband and his girlfriend might be sending each other Facebook gifts such as "Love Birds" and posting about the need for discretion. (Husband: "[n]o more Facebook. . . to public for me." Girlfriend: "LOL o.k. under the radar . . . flying low. . . "[5])

In one shocking incident, a woman learned on Facebook that her current husband had just tied the knot with someone else. In a beautiful seaside setting in Amalfi, Italy, Lynn married John France, the man of her dreams. The photos of their wedding were so radiant that the company that arranged the ceremony uses one on its website. The couple had two beautiful children, a home in Ohio, and— Lynn thought initially—a wonderful marriage.

Then, on Facebook, Lynn glimpsed at a photo of another woman's dream wedding on the woman's public page.[6] The ceremony took place at Disneyworld. The woman was dressed as a princess, and her new husband was dressed as a prince. But that woman's dream wedding was Lynn's nightmare. The prince in the photo was Lynn's husband, John, who then moved in with his new wife. When John visited his children back in Ohio, he put them in the car and drove them to his Princess Bride in Florida. Lynn's children started appearing on the other woman's Facebook page and Lynn began an arduous legal process to attempt to regain custody. Lynn filed suit for divorce, but John claimed that, because of a technical glitch, their Italian marriage wasn't legal. So, he said, he wasn't a bigamist for marrying another woman. The company that organized Lynn and John's Italian wedding disagreed; it issued a statement saying, "The legal wedding in Italy is valid all over the world. We would like to understand better why someone is lying about this."[7] Lynn's attorney pointed out that if the marriage was not valid, then John, who filed joint tax returns with Lynn, had "lied to the IRS . . . insurance companies. Banks."[8]

Divorce lawyers have also used photos from social networks to find out about assets one spouse is hiding from another. In one New York case, the husband had no job and claimed he couldn't afford child support. But a search of Myspace found him listed under a different name as a co-owner of a hot Bronx club.[9] When a 38-year-old man thought his estranged wife had hidden income sources, investigators he'd hired turned straight to social networks. They found pictures on her Facebook profile of her in Aruba and dining at the Ritz-Carlton in Maui. By following her Twitter posts, the investigators learned that her family owned a successful restaurant and that she was involved in expanding the business, which paved the way for them to uncover her hidden wealth.[10]

The most common—and most troubling—use of social network information is in the context of custody. Snap judgments are being made by judges based on postings and tagging—reflections of the person's second self. Digital data is used in divorce and child custody cases without a context. People are losing their rights to see their children because of ill-advised posts about their thoughts or their lifestyle, like posts implying they are single when they are married or posts suggesting they approve of smoking pot. In many instances, the judges aren't seeking additional evidence to see if the child is in danger. But does poor judgment about what you post mean that you would necessarily be a bad parent?

When my son was a month old, propped up in a stroller, a friend of mine unsnapped the first two buttons of his onesy and put an Elvis-style gold chain around his neck. Then she put a sealed bottle of Jack Daniel's in the stroller next to him. She shot a photo of him and then removed the necklace and the bottle. It was a joke, but what if she'd posted it on a social network? The Department of Children and Family Services might have tried to take my baby from me. When a Florida woman posted a photo of her baby next to a bong because she thought it was humorous, she was investigated by the state.[11] Her parental rights weren't terminated because neither the bong nor the baby were found to have drugs in them. But, if she were involved in a divorce dispute, her husband might have used the photo to convince the court to deny her custody.

Since our Social Network Constitution includes a right to connect, freedom of expression, and a right to privacy, we've got to consider how these rights should be interpreted in the context of the family. Should people's social network postings and photos be used against them in custody cases? Or should other core rights override that possibility?

Most people don't realize how fiercely the U.S. Constitution protects the rights of parents, married or not, to maintain a relationship with their children. Considered part of the rights of liberty and privacy, a person's right to bear and raise children is considered to be a core element of our society. The parent/child relationship cannot be terminated unless a parent abandoned or neglected the child or was an unfit parent. Parenthood is considered, under the U.S. Constitution, as one of the "basic civil rights of man."[12] That right is described by the Supreme Court as "far more precious than property rights."[13]

Beyond the basic right to retain a connection to his or her child, a parent has a right to determine how to raise that child. The U.S. Supreme Court has intervened when the federal or state government has improperly constrained how a parent raises a child. A century ago, parents challenged state laws prohibiting public and private schools from teaching foreign languages. More recently, in the 1970s, the laws requiring compulsory education until age 16 were challenged by Amish parents, who educated their children at home after middle school.[14] In both those situations, the Supreme Court weighed in on the side of the parents to make the primary decisions about how to rear their children: "It is cardinal with us that the custody, care and nurture of the child reside first in the parents, whose primary function and freedom include preparation for obligations the state can neither supply nor hinder."[15]

The Supreme Court, in tackling the law against teaching children a foreign language in school, referred to the Greek philosopher Plato's idea that children be raised communally, never knowing their parents, and the ancient Spartan proposal that the state take over the rearing of boys. The Court said, "Although such measures have been deliberately approved by men of great genius, their ideas touching the relation between individual and state were wholly different from those upon which our institutions rest; and it hardly will be affirmed that any Legislature could impose such restrictions upon the people of a state without doing violence to both letter and spirit of the Constitution."[16] The Supreme Court affirmed parents' leeway in making decisions about their children, saying "The fundamental theory of liberty upon which all governments in this Union repose excludes any general power of the State to standardize its children."

Despite those core protections of parenthood, courts have taken some missteps over the years when faced, not with lofty analysis of Constitutional rights, but the down-and-dirty decisions in bitterly contested divorce cases. Sometimes stereotypes, prejudice, and improper social norms have provided the basis for wrenching custody from one parent or the other. And social networks have magnified that problem.

Until the start of the twentieth century, children were viewed as the property of their fathers, who gained custody of them if the parents separated. Then courts developed the "tender years" doctrine, creating a presumption that young children were better off being reared by their mothers. The "tender years" doctrine was completely abolished by the 1970s, when ideas of equality influenced courts to give equal weight to each parent's interests, with the goal being to determine the "best interests of the child."

But even the noble best interest standard has been applied in prejudicial ways. Only in recent years have courts been willing to say that a person's religion, sexual orientation, age, or low income shouldn't be grounds for giving the child to the other parent. But now, social networks have provided the means for prejudice again to guide whether a parent will be able to retain the right to raise his or her child.

As in the employment context, people whose Facebook photos show them engaging in otherwise legal behaviors—like drinking—are finding that their parental fitness is being challenged. Sometimes there are extenuating circumstances. The parent might be a minor, or perhaps the parent's therapist warned against drinking. But should that really be enough to limit a parent's rights?

Decisions about custody of a child or termination of parental rights should legitimately be based on the best interests of the child. This can involve an assessment of how a parent treats the child, whether the home environment is a safe and nurturing one, whether the custodial parent will foster a good relationship between the child and the other parent, and, if the child is old enough, the child's preference.

But using all of a parent's social network postings to make those decisions is not appropriate. Social network posts and photos should be used only if they are directly relevant to the best interests of the child. Otherwise, they might unduly prejudice the judge against the parent and lead to an inappropriate denial of parental rights.

The problem is compounded because social networks reveal aspects of a parent's life that judges have not had access to before. In theory, that might sound like a good thing. But there is a huge class divide between the (mostly white, mostly older, mostly well-off) judges who oversee custody cases and the parents who come before them. Judges can now glimpse into the lifestyles of people quite different from themselves, and may be repelled by aspects that have nothing to do with whether the person whose Facebook or Myspace page they are viewing would be a good parent or not. People are revealing their most intimate selves in their postings. And some of what they reveal may be more prejudicial than probative if admitted in the case.

Judges have gone so far overboard in admitting digital evidence that some have granted a spouse access to the other spouse's entire hard drive. In a New York case, the husband had been issued a laptop as part of his job at Citibank. He would occasionally let his children do their homework on it. When the wife took the laptop to her lawyer's office, a judge had no problem issuing an order that allowed her to copy the entire hard drive. Citibank intervened, saying that the computer belonged to the corporation and not the husband. The court held that because his children had used the computer, it was controlled by the husband, not his employer. The judge said the laptop was no different than an unlocked filing cabinet in the family home.[17]

The only way to guarantee that your posts won't come back to haunt you in a custody case would be never to have had a social network page or to act like a Stepford parent and post only positive and glowing things about your every moment with your child. (Perhaps even doing that would backfire since it could be used to show that you are too enmeshed in your child's life and won't give your child enough space to grow.) Erasing a page you've previously created or deleting your social network presence entirely won't help. Projects such as the Wayback Machine have probably captured screenshots of that page in its earlier incarnation.

Since parenthood is rewarding, demanding, and frustrating all at the same time, people may unthinkingly blurt out their frustrations in social media. What if you once tweeted that you didn't want children? Should that statement be used to terminate your parental rights? In *In re T.T.*, a Texas case, the court allowed such a statement from a dad's Myspace profile to be used against him.[18] What if you failed to mention kids on your Match.com profile? Would that show you were a bad mom? How about if you said, "I love my motorcycle" or "I love my iMac" but didn't mention your children? Would that indicate that your kids played second fiddle to your possessions? The fundamental protection of the parent/child relationship should mean that any statements about the child be kept out of the case unless they indicate that the parent is likely to harm the child emotionally or physically. And lack of statements about the child (or even a statement that one doesn't have children) shouldn't be used as a way to show parental unfitness.

Likewise, courts should not assign custody based on what a parent sets as his or her emotional status on Facebook or Myspace—or on general statements about anger or depression. When a mother vying for custody claimed that her husband had a bad temper, he was faced on the stand with a printout of his own Facebook page: "If you have the balls to get in my face, I'll kick your ass into submission."[19] Does that necessarily indicate that the person will abuse his child—or might it just show the lengths he'll go to protect him?

What about information about the parent's lifestyle? Does a woman's tattoo of 666 in a Facebook photo indicate she might subject her child to a Satanic cult? When a deadbeat dad talked about how all he needed was alcohol, women, and sticky dank (pot) on his Myspace page, a court said, "his Myspace page further suggests that his lifestyle is not conducive to one in the best interest of a child."[20] The mother didn't actually have to prove he'd bought or smoked marijuana before she was able to terminate his parental rights.

In another case, the dad was supposed to be watching the child, but a YouTube video that was posted later showed he was at a party.[21] Should that incriminating video be allowed into evidence in a custody case, or should it be kept out if he can show that he'd arranged for someone else to watch the child safely?

Women who express their sexuality on social networks run the risk of losing their children entirely. Judges seem to have a stereotypical notion that women can't be good mothers if they think about or engage in sex. A woman who posted sexually explicit comments on her boyfriend's Myspace page lost custody of her child as a result.[22] Another woman, who was described as portraying herself as a soccer mom, lost custody when she posted sexually explicit photos of herself seeking hookups.[23] In another case, the court considered the mother's Myspace photographs of herself with different men, as well as photos in which she was surrounded by individuals who were intoxicated—and granted custody of the child to the father.[24] Sometimes judges consider it bad parenting if the mom lets the child see something provocative on her social network page. When a woman created Myspace pages for her seven- and ten-year-old daughters and friended them, they could then view her Myspace page, which contained photos that the judge said showed her posing "provocatively" in lingerie.[25] The judge awarded custody to their father.

Kathy and Robert Lipps married in September 2007. Four months later, his National Guard unit was called to active duty in Iraq. Kathy found herself not only alone, but pregnant. As the pregnancy progressed, she got involved with another man. Her husband learned of the affair when he saw Kathy's Myspace photo of the two of them kissing.[26]

Granted, cheating on someone while he is far away fighting for his country is a horrible thing. But lots of relationships break up, for all sorts of reasons, distance among them. Does falling in love with someone else mean that you will necessarily be a bad mom?

When the couple's baby was born, Robert came home on leave and then left again. Kathy moved in with her new boyfriend and his mother, who helped care for the child. When his tour of duty ended a few months later, Robert filed for divorce and sought emergency custody of his son. The judge seemed enraged that a pregnant woman had gotten involved with a new man and even posted it on Myspace. The male judge told her, "I don't know where your mom and dad went wrong, and I don't think they did, and I guess this is a nightmare every parent faces when you have got a child that acts like a slut, quite frankly, a slut."[27] The judge granted the husband a divorce and awarded him custody of the child, even though he'd not seen or cared for the child except for during his brief leave. Now the ex-husband has moved the child to another state. Kathy, on the advice of her lawyer and the judge, stopped seeing her boyfriend. She says, "no man is worth losing my son over." But that is exactly what happened, due to the Myspace photo.

Can you imagine that ever happening to a man? Let's assume that Arnold Schwarzenegger did impregnate his maid in the family home when his wife was pregnant. Is there any chance that a male judge would call him a slut and take his four children away from him? What about former Congressman Anthony Weiner's X-rated tweets? Will that mean that the judge won't allow him the right to a relationship with the baby his pregnant wife was carrying at the time of the tweets?

In divorce cases, people are angry. They may post negative comments about their exes on their social network page or even on the walls of their friends' pages. Yet should those emotional outbursts be used to indicate that they wouldn't play fair with the other spouse if they were given joint custody? Some courts are willing to sever a person's custody rights if there is even a hint that the person might not share visitation or parenting. A court in South Carolina awarded sole custody to a mother in part based on this post on the father's Myspace page: "I'm actually a little sorry for [Mother] just because losing her job affects my children. Well, maybe not. Now I'm financially more than able to support them if [Mother] gets out of the way or is pushed out. I can provide them the life they deserve and even better that they lost . . . I have a wonderful life planned with a wonderful woman and my only goal now is making sure my children share it with us." The court felt that the mother would be more likely to foster a good relationship between the children and her ex than the dad would.[28] But there was no evidence the dad had shown the page to his daughters or was planning to keep the children from seeing their mother. And who's to say that the mother wasn't poisoning the kids against their dad by what she said, even if she didn't post it on Myspace?

When the child is older, his or her own Facebook and Myspace posts might be used to determine custody. In a 2010 Wisconsin case, a sexy photo of their daughter on Myspace led to the transfer of custody of the girl to her father on the grounds that her mother wasn't properly supervising her.[29] But should a teenager's inappropriate post be held against the custodial parent? In the age of Paris Hilton and "sexting," couldn't this happen to any parent?

Some courts are beginning to limit the use of social network information, and that seems appropriate. A few years after her husband received full custody of their two young daughters, Elizabeth Gainey asked the court to reconsider the arrangement. In addition to introducing evidence of how the girls' health had deteriorated while with her ex, Elizabeth tried to introduce evidence from the Myspace and other internet accounts he and his new wife had created. She wanted the judge to consider photos and links they'd posted on social networks, including a picture of the new wife in a French maid outfit, a movie trailer featuring a naked woman sold into slavery, a picture of Charles Manson, and a video of Ronald McDonald being shot in the face.[30] When the judge refused to admit the evidence from Myspace, Elizabeth appealed, arguing that his sexually explicit and violent material indicated that he was an unfit father. The appellate court ruled that the trial court was well within his rights not to admit the posts from Myspace but did send it back for an exploration of whether the girls' health issues were enough to reassign custody. That decision seems right—that the focus should be on the condition of the children, not the dreck that adults include on their Myspace page.

That's not to say that social network posts should never be admitted in custody cases. But generally, even when they indicate the possibility of parental unfitness, their best use is to spur further investigation, not to serve as the sole reason to strip parents of their rights. One woman posted that she did drugs, but only when her child was asleep. Should her child be taken away—or should she at least be subject to urine testing for drugs so that courts can see whether that is true and whether it is a continuing problem?

More troublesome is an instance where an ex-wife was living with a man who'd abused children in the past. On her Myspace page, she mocked her own children's allegations of abuse by the man.[31] Unlike so many of the other instances where posts unrelated to parenting are used to sever parental rights, here there is a potential direct relationship between the mother's post and her role as a parent. That should be a trigger for immediate intervention and concern.

Judges who put a high value on social network posts aren't necessarily acting in the best interests of the children. They may be awarding custody not to the best parent but to the one who is most technologically challenged and thus doesn't have a Facebook or Myspace page in the first place or the one who games the system by posting sweet but untrue posts.

The focus on social network information also pushes parents to spy on each other or hire expensive experts to break into the other spouse's computer. This ignores the rights of other parties—like the employer of the spouse—who might have an interest in the computer's contents and, as in the Citibank case, may actually own the computer. Plus, it can't be in the best interests of the child if Mom or Dad ends up in jail as a result of hacking the other parent's Facebook page.

In Florida, Beverly Ann O'Brien secretly installed spyware on her husband Kevin's computer. At certain intervals, the program captured shots of what the husband saw on his screen, including messages, emails, and websites. The court refused to allow the screenshots to be used as evidence, saying that the wife had violated a Florida wiretap statute that prohibited the interception of "wire, oral or electronic communications." Not chagrined by being busted for wiretapping, she argued that the statute didn't apply because the program she installed didn't "intercept" the emails but rather copied them and sent them to the hard drive once they hit the screen. The court did not agree. The judges said, "The Wife argues that the communications were in fact stored before acquisition because once the text image became visible on the screen, the communication was no longer in transit and, therefore, not subject to intercept. We disagree. We do not believe that this evanescent time period is sufficient to transform acquisition of the communications from a contemporaneous interception to retrieval from electronic storage."[32] Because they'd been obtained illegally, Beverly couldn't use them as evidence against her husband.

In contrast, a New York judge bought the argument that the wife hadn't engaged in wiretapping when she copied emails from her husband's computer since they weren't intercepted en route; she was able to use the emails against her husband in the divorce case.[33] In an Arkansas case, the husband was found not to have violated the federal Wiretap Act when he installed a computer program on his wife's computer to copy her keystrokes, which allowed him to obtain his wife's passwords.[34] But he was held to have violated the federal Stored Communications Act and the Arkansas computer trespass law.

Some ex-spouses resort to even more elaborate ruses. Angela Voelkert was engaged in an acrimonious child custody dispute with her ex-husband, David. She created a fake Facebook identity, pretending to be a 17-year-old girl, "Jessica Studebaker," and friended David to glean damaging information she could use against him at trial.[35] In 2011, Angela asked a court for a restraining order against David and, as support, attached several pages of Facebook printouts, where David said he'd put a tracking device on his ex-wife's car[36] and also said, "Once she is gone, I don't have to hide with my kids. I can do what I want and not have to worry about not seeing my family anymore. You should find someone at your school. There should be some gang-bangers there that would put a cap in her ass for $10,000. I am done with her crap!"[37]

Police arrested David for allegedly planting an illegal GPS device in his ex-wife's car[38] and started an investigation into his plan to kill his ex. He spent four days in custody until he convinced prosecutors that he'd known that his ex-wife was behind the fictitious profile and was just playing along to ensnare her. He produced a sworn, notarized affidavit written before he'd responded to Jessica. The document described receiving a friend request from "Jessica Studebaker," whom David Voelkert suspected was his ex. The affidavit went on to state, "I am lying to this person to gain positive proof that it is indeed my ex-wife trying to again tamper in my life. . . . In no way do I have plans to leave with my children or do any harm to Angela Dawn Voelkert or anyone else."[39] David Voelkert had kept one copy and given another to a relative for safekeeping.

In another case, an ex-husband's new wife created a fake account and sent threatening messages to herself, pretending to be the ex-wife. When the ex-husband tried to introduce them in court to challenge his ex-wife's custody, an investigation revealed that the ex-wife hadn't sent them. But, in other cases where a divorcing spouse doesn't have the resources for computer forensics to prove that she hadn't sent such messages, a judge might admit the posts and use them to deny an innocent parent of custody.

Even though divorce courts have found spouses culpable for trying to dig up dirt—or even to manufacture it—family law judges routinely grant orders that require husbands and wives to turn over all their social network postings and sometimes their entire hard drives. Those same judges are using posts on social networks to make moral and legal judgments about people, including whether they should be allowed to retain any rights to their children.

How should our Social Network Constitution address the use of social network postings in cases that involve one of our most important rights, the right to bear and raise children? People post plenty of stupid stuff on social networks. But should a foolish slip be enough to force you to give up your child? Judges in child custody cases have a huge amount of discretion. The standard is a loose one—"the best interests of the child"—and there is no jury present to layer in some understanding of the community standards on social network behavior.

What principles might our Social Network Constitution provide to guide judges in family disputes, particularly with respect to custody? The privacy and liberty rights of parents, plus their right to a fair trial, should weigh against the use of social network posts in determining custody unless the posts signal an imminent risk to a particular child.

Under our Social Network Constitution there should be tighter constraints on what evidence gets admitted from social networks and what assumptions are made based on that evidence. If social network postings were protected by fundamental rights and rarely allowed to be sought or introduced as evidence, custody decisions could then be based on how the parent actually treats the child, rather than the level of bravado or seduction with which the parent addresses the digital world.

11

Social Networks and the Judicial System

Rufus Sims faced an uphill battle in Judge Shirley Strickland Saffold's courtroom. His client wasn't exactly a sympathetic defendant. Anthony Sowell, a registered sex offender, stood accused of mass murder and faced the death penalty.[1] Police had found two dead bodies on the third floor of his house, another dead body under the stairs, and two in the crawl space.[2] Five more bodies were buried in the yard, along with a skull in a bucket in the basement.[3] A year earlier, Sowell had been accused of rape in that same house.[4] A bloody, naked woman had flagged down police and said he'd dragged her into the house and tried to rape her. She'd escaped by falling through a window. Police found her bloody clothes in Sowell's trash can. He'd been arrested, but charges were dropped since the police didn't find the victim to be credible.[5] Neither the police nor parole officers had searched the home.[6] "Nobody knew this guy was a predator—the system dropped the ball," his neighbor Raymond Cash, Jr., told ABC News.[7]

Rufus had navigated tough cases before and was used to reading heinous accusations about his clients—some true and others not—in the media and in blogs. But a posting on the Cleveland *Plain Dealer* website in November 2009 was something altogether different. The target of the post was not a client but Rufus himself. An anonymous poster with the screen name of "lawmiss" commented on another case Rufus had handled in Saffold's court, involving Angela Williams, a city bus driver who killed a pedestrian in a crosswalk.[8] Even though Sims had gotten the charge reduced to vehicular manslaughter, with just a six months' jail sentence, lawmiss had posted, "Rufus Sims did a disservice to his client. If only he could shut his Amos and Andy style mouth. What makes him think that is [*sic*] he insults and acts like buffon [*sic*] that it will cause the judge to think and see it his way. There are so many lawyers that could've done a much better job. This was not a tough case, folks. She should've hired a lawyer with the experience to truly handle her needs. Amos and Andy, shuffling around did not do it."[9]

Who had done this to him? Someone in Angela Williams's circle of family or friends? Someone he'd turned down as a client? A former friend he'd offended?

Despite his legal training, Rufus couldn't do anything to get the post removed. Under the federal statute 47 United States Code Section 230, the website couldn't be forced to remove it and was not liable even if the comment was found to be defamatory, racist, or untrue. And there was no avenue to get the name of the anonymous poster.

Protection of anonymous posters does make a certain amount of sense. If free speech about public matters is to be encouraged, especially about something as important as the legal system, anonymity might be warranted. But, for a robust discussion, doesn't the identity of the poster matter? How can someone judge the credibility of a comment if he doesn't know who was behind it?

Lawmiss was a frequent visitor to the newspaper's website. She posted more than 80 comments about court cases, about sports, and even about the relatives of reporters. One post, commenting on the mental state of a relative of James Ewinger, a *Plain Dealer* reporter,[10] upset an editor of the paper. Using the software that posted stories to the website, the editor looked up lawmiss's email address. He then linked the address to a person through a simple Google search.[11]

Lawmiss was an email address used by Judge Shirley Strickland Saffold. And several of the posts involved cases that she was presiding over.

Rufus Sims was infuriated when he learned that the account criticizing his "Amos and Andy style mouth" had been linked to the judge.[12] He moved to recuse her from the Sowell case. But she refused to budge. She claimed that her daughter, not her, had posted the comments. And she turned around and sued the newspaper for $50 million for invading her privacy and that of her daughter.

Judge Saffold and the newspaper already had a rocky relationship. During her 16 years on the Court of Common Pleas, she'd criticized the paper's coverage of her court. In 1996, the newspaper's court reporter, James Ewinger, had reported on Judge Saffold's advice to a woman who pled guilty to credit card fraud: Find a man.

"Men are easy," the judge told the defendant. "You can go sit at the bus stop, put on a short skirt, cross your legs and pick up 25. Ten of them will give you their money. It's the truth." Saffold also said, "If you don't pick up the first 10, then all you got to do is open your legs a little bit and cross them at the bottom and then they'll stop."[13]

Further investigation of the lawmiss posts found that both the judge and her daughter used that email account. Some of the posts could have come from either of them—like the October 2009 comment about model Heidi Klum deciding to take her husband Seal's last name.[14] But a public records request by the newspaper indicated that some of the comments had come from the judge's court computer.[15]

Rufus Sims was worried about the Sowell case. "This shows a personal disdain for me and a personal bias against me that she could easily take out on our client," he said. He didn't believe that Saffold's daughter was behind the rant. "That doesn't make any sense to me. Someone else is using the judge's account? Come on. Why would Sydney do it? I don't get it."

Sydney Saffold, a 23-year-old former law student, took the heat for all the posts.[16] But, even if she'd been the one to criticize Rufus, how had she gotten the information? Had she sat in the courtroom or heard it from her mother? And even if the criticism was Sydney's alone, it still seemed improper for the judge to stay on the Sowell trial. Wouldn't there be some pressure on the judge to rule against Rufus, just to prove her daughter right? And if there were 80 posts by lawmiss,[17] why had Sydney, when asked about the frequency of her postings, said, "quite a few, more than five."[18]

While Rufus was pondering his options, Judge Saffold was continuing her suit against *The Plain Dealer* and Advance Internet, the creator of the newspaper's website, claiming that they breached the website's privacy policy. Saffold told the newspaper, "This smut and smear crap you're doing is disturbing."[19]

The newspaper's editor, Susan Goldberg, argued that the paper's actions were appropriate. "What if it ever came to light that someone using the email of a sitting judge made comments on a public website about cases she was hearing, and we did not disclose it?" Goldberg asked. "These are capital crimes and life-and-death issues for these defendants. I think not to disclose this would be a violation of our mission and damaging to our credibility as a news organization."

Goldberg's claim that disclosure was in the public interest seemed compelling, but Goldberg and the newspaper caught flak from other posters on the paper's website, who felt it was crucial to maintain anonymity.

Ultimately, Judge Saffold dropped the lawsuit against *The Plain Dealer* and settled with Advance Internet for an undisclosed sum.[20] Advance Internet now blocks newspaper employees from accessing the email addresses of people who post.[21]

Eventually Rufus Sims did get the judge removed from the Sowell case.[22] Ohio Chief Justice Paul E. Pfeifer ordered her off the case, noting that an objective person might reasonably question why a judge's personal account would contain posts with personal opinions about lawyers litigating before her.[23] Judge Saffold continues to deny that she made the comments. "The fact that my computer was open to Cleveland.com at the same time that lawmiss posted comments is merely coincidence—two things happening at the same time, but with no causal relationship between them," she said.[24]

The judicial system is a precarious institution in any society. The system requires a huge buy-in. People have to believe in the rules of the endeavor. The black robes, the call of "Oyez, oyez" at the beginning of a courtroom session, the carved wood and marble of a typical courthouse—all aspects of the proceedings are designed to command respect for the institution. Who would agree to abide by the decision of a judge dressed in a Hawaiian shirt and a baseball cap?

The need for respect is why the ethics codes for lawyers and judges require that they "avoid the appearance of impropriety." Under their canon of ethics, judges also must not make it appear that they are in a position to improperly influence the outcome of the case.

Jurors, too, are bound by certain rules. They have to wrap their heads around the idea that the legal process is not about truth per se, but a judgment based on admissible evidence. If evidence was obtained impermissibly (for example, through a search without a warrant), it cannot be considered. Evidence that is unrelated to the case and may be prejudicial to the defendant is not allowed to come before the jury. Before deliberation, the judge instructs jurors on how they are supposed to interpret the terms of a statute or the legal responsibilities of the parties. Judges and lawyers must maintain confidentiality about their cases. Breaches of these rules can mean removal from a case or even disbarment.

Jurors aren't supposed to be swayed by emotions unrelated to the case (such as their own past dealings with the defendant) or by anything they've read in the news media or any other source. They are not allowed to talk about the case with others, so that they don't base their decision on the influence of a friend or relative, rather than their own judgment. They can't visit the scene of the crime on their own or conduct their own investigations.

Before social networks and other internet developments, it was pretty easy to keep jurors in line. If a case garnered a lot of attention in the local press, a judge could order a change of venue so that the case was heard in another town. If the court was worried about outside influences, the jurors could be sequestered.

But now court cases, like every other aspect of our lives, are being reshaped by search engines and social networks. With most Americans on social networks, posts and tweets are creating new challenges for the fundamental right to a fair trial. How would you feel if you were on trial for your life, wrongly accused, and the judge was tweeting during trial and not paying attention to you? How about if the prosecutor "friended" a juror and started having sex with her? And what if a juror posted the facts of your case on her Facebook page and asked her friends to vote about whether you should be declared innocent or not? Such incidents have already occurred as social networks invade the criminal justice process.

In Georgia in 2009, a 54-year-old judge, Ernest "Bucky" Woods III, contacted an attractive 35-year-old woman, Tara Black, through Facebook.[25] This would have been a routine activity but for the fact that Tara was a defendant in his court, on trial for theft by deception.[26] She wrote back, asking him for help. She'd unthinkingly posted a photo of a friend of hers holding a beer, which violated the terms of his probation.[27] Her Facebook gaffe had sent him back to jail. Could the judge intervene on the man's behalf? She told the judge she'd have a female friend give him a year's worth of free massages in return for his help, then added that it was a joke, "LOL," and "I'm not really trying to bribe you."

The judge refused to help the probationer but did offer Tara advice on her own case. According to a *Fulton County Daily Report* analysis of 33 pages of correspondence between the judge and the defendant, he signed an order releasing her on her own recognizance without having to post a cash bond. He told her he'd convinced the district attorney to defer her prosecution so that she could raise the money to pay off her debt and told her he'd dismiss the case once the money was paid. He cautioned her not to talk to anyone about the arrangement. "I can help you more behind the scenes so long as very few people know it," he wrote.[28]

The emails also indicated that he had visited her apartment and helped her out with her rent. "OK. I am going to be really brave and ask you a favor," she apparently wrote to him, asking him to loan her $700 for an apartment. He asked if $450 would help. She answered, "OMG! Yes."

When the emails came to light, Judge Woods claimed they were fake, then recanted and said that only some were fake but didn't indicate which ones.[29] Instead, with an investigation of his judicial conduct imminent, he stepped down. Woods was already fully vested in the state's pension plan.

"I call it a retirement," Woods told the *Fulton County Daily Report.* "I just got tired of living under a microscope."[30]

Under a microscope? He wasn't exactly Brad Pitt, a celebrity followed around by paparazzi. He was a judge who, even if he'd just sent a single Facebook message to a defendant, had crossed ethical boundaries. But so much of modern life occurs on social networks that participants in all areas of the judicial system are violating moral and legal strictures in ways that create Constitutional and ethical challenges.

Over 40% of judges use social media.[31] But judges are usually older than other participants in the legal system and, Bucky notwithstanding, are concerned about not giving the appearance of impropriety (a mandate of judges' ethics codes). Other participants in the judicial system, such as the attorneys who appear before them, are younger and more used to living their lives online. While only 15% of attorneys had a social network presence in 2008, more than 56% had social network profiles two years later.[32]

In Illinois, an assistant public defender, Kristine Peshek, published a blog, "The Bardd Before the Bar—Irreverant Adventures in Life, Law, and Indigent Defense."[33] Her blog was open to the public and was not password-protected.[34] In her posts, she referred to her clients by their first names, a derivative of their first names, or their jail identification numbers.[35] Peshek described a client "taking the rap for his drug-dealing dirtbag of an older brother," referred to another as "standing there in court stoned," called a judge "a total asshole," and referred to another jurist as "Judge Clueless."[36]

In an April 2008 blog post, Peshek said her client, who was accused of forging a prescription for a painkiller called Ultram, had claimed that she wasn't using any drugs at the time of her sentencing. But as they were leaving court, the client revealed her methadone use to Peshek. In her post, Peshek wrote: "Huh? You want to go back and tell the judge that you lied to him, you lied to the pre-sentence investigator, you lied to me?" As a result of the post, Peshek was accused of several ethics violations: failing to ask a client to rectify a fraud on the court; failing to disclose to a tribunal a material fact known to the lawyer when disclosure is necessary to avoid assisting a criminal or fraudulent act by the client; conduct involving dishonesty, fraud, deceit or misrepresentation; conduct that is prejudicial to the administration of justice; and conduct that tends to defeat the administration of justice or to bring the courts or the legal profession into disrepute.[37] Peshek lost her job, and on May 18, 2010, she was suspended from the practice of law for 60 days by the Illinois Attorney Registration and Disciplinary Commission.[38]

Judges and lawyers aren't the only participants in the judicial system whose reliance on social networks has come back to haunt them. Some people are so dependent on social networks that they can't make a decision about anything—whether to buy a certain car or break up with a boyfriend—without doing internet searches or running a poll of their friends. When faced with the evidence in a British sexual assault and abduction case, a juror posted the facts on her Facebook page and said, "I don't know which way to go, so I'm holding a poll."[39] A number of her friends wrote back, urging a guilty verdict. As in the United States, Great Britain's Constitution guarantees a right to a fair trial, which limits juror decision making to what occurs in the courtroom. The juror was removed from the case, and the other 11 decided to acquit.[40]

Jurors' use of Facebook, Twitter, and Google have led to dozens of mistrials and overturned verdicts. In 2009, in a single court, 600 potential jurors were dismissed when prospective jurors mentioned they'd done research about the case and discussed it with others in the jury pool.[41]

With the click of a mouse or a simple search on their smartphone, jurors can find out the skinny on the lawyers, seek out past misdeeds of the defendant, make assessments of the credibility of witnesses, and even use Google Earth to revisit the crime scene. But all of these actions violate a defendant's right to a fair trial. Web searches uncover information from outside of the courtroom, where there is no chance for the defendant's lawyer to correct inaccuracies or cross-examine the individual who provided the information.

Social networks and search engines might also lead jurors to groups that deliberately provide false information or use public relations techniques to sway the jurors. Attorney Doron Weinberg says that during Phil Spector's murder trial, a blogger was posting untrue and damning information, including a reference to an alleged confession Spector had made.[42] During Martha Stewart's securities fraud trial, the consummate perfectionist maintained a website, Marthatalks.com, in an attempt to influence public opinion. The site was updated almost daily with trial updates, supportive letters and newspaper editorials, and Stewart's own statements of her innocence. Launched almost immediately following her indictment, Martha talks.com was viewed by 16 million people in the first six months of its existence.[43]

With so much reliance on social networks to find a restaurant, a job or a lover, it's hard to convince jurors that they shouldn't similarly search for information related to a case. In fact, many erring jurors have felt they were merely carrying out their responsibilities by getting additional information.

When a Pennsylvania high school librarian, Gretchen Black, was called to jury duty on a case involving a man who allegedly shook his girlfriend's one-year-old baby to death, she handled the situation as she did everything else in her life.[44] She did research.

The defendant had allegedly shaken the baby girl so hard that she suffered a brain injury, retinal detachment, and a fractured skull.[45] Gretchen, Juror Number 11, went online to find out more about retinal detachment. During deliberations, the jury decided to find the defendant not guilty of first-degree murder. When they turned to the lesser offenses, involuntary manslaughter and third-degree murder, Gretchen offered to describe what she'd found online. Instead, the jury forewoman told the judge about the breach. The result: a retrial on the lesser offenses and the threat of criminal sanctions against Gretchen herself.

Luzerne County Assistant District Attorney Michael Vough sought contempt charges against Gretchen, pointing out that the judge had repeatedly warned jurors against going online to conduct research. The judge appointed an attorney to represent Gretchen. "She just wanted to be the best juror possible," Gretchen's attorney told Reuters. Early in the case, she'd raised her hand and asked if she could question a prosecution witness, which, of course, was not permitted. Gretchen's attorney argued that she knew she wasn't supposed to undertake online searches about the case itself but didn't realize the prohibition included more general research on related issues.

Jurors like Gretchen have tainted cases when they've sought information about a witness, a defendant, or a concept at issue in the case. In sexual assault and incest cases, jurors have looked up Myspace and Facebook profiles of alleged victims and made assumptions about their credibility based on their photos and posts. In rape cases, courts are not supposed to admit evidence of the previous sexual history of the victim, but jurors have concluded that the victim is a slut based on how she looks in her social network photos.

In one case, a behavioral specialist took the stand and testified that the incest victim suffered from oppositional defiant disorder (ODD), though the specialist did not explain the disorder.[46] A juror looked it up and found some information that ODD was associated with lying. It's easy to see how that information could lead jurors to doubt the victim. And since the information was learned outside the courtroom, there's no chance to question its validity. A higher court overturned the conviction, holding that once the juror's actions came to light, the judge should have specifically asked each juror if he or she could continue to be impartial.[47]

In other cases, jurors' transgressions involve information about the defendant. When a truck driven by Clyde Sharpless snagged a cable suspended from a telephone pole, the pole crashed into Dong Sim's car, killing her husband and son. The jury found Clyde liable.

Then it came to light that a juror had conducted independent internet research on Clyde's previous driving record during the trial. Clyde appealed, but the appellate court held there was no harm. The website the juror consulted showed only that Clyde had some prior traffic violations, and the juror did not recall viewing any information pertaining to Clyde's drug or alcohol use. The juror testified that the information she'd viewed had no impact on her deliberations and that she hadn't shared the information with any other juror.[48] But shouldn't there be a bright-line rule? Can we be confident that, when confronted by a judge, a juror would disclose not only that she'd breached the judge's order, but that her breach led her to decide unfairly against a defendant?

When a juror used his iPhone to look up the word "prudence," a key legal concept in a manslaughter trial, he discussed what he found with the other jurors.[49] After the defendant was convicted, the slip came to light. An appellate court granted the man a new trial, saying "Although here we confront new frontiers in technology, that being the instant access to a dictionary by a smartphone, the conduct complained of by the appellant is not at all novel or unusual. It has been a long-standing rule of law that jurors should not consider external information outside the presence of the defendant, the state, and the trial court."[50]

Jurors' social network misconduct involves not only improper input, but also improper output, such as tweeting or posting material on Facebook. In Barry Bonds's appeal in his prosecution for lying about using steroids, his former girlfriend testified about the changes in his body while he was with her, purportedly due to steroid use. His attorney, Cristina Arguedas, obtained a no-tweet order for the jurors since she was worried they would disclose salacious material and do online research about the case. The judge told the jurors not to discuss the case "in person, in writing, by phone or electronic means, via email, text messaging, or any Internet chat room, blog, website or other feature."[51]

If a bystander at a trial or a reporter can tweet about a witness's testimony, why can't a juror? The concern is that a juror's tweet will distort the trial. What if, for example, one side's attorney read the tweet and introduced material in the case just to respond to that juror? What if the juror could get a bigger audience by exaggerating or distorting what was happening in court? And what if people responding to the tweet give the juror information about the defendant or the victim that he wasn't supposed to know?

In 2010, Reuters monitored tweets over a three-week period for the term "jury duty." The study found that tweets from jurors or prospective jurors pop up at the rate of one every three minutes.[52] Ignoring their legal duty, some jurors make up their mind before all the evidence is presented. "Looking forward to a not guilty verdict regardless of evidence," one person tweeted. Another said, "Jury duty is a blow. I've already made up my mind. He's guilty. LOL." Yet another man, in a jury pool, hadn't even been selected for the trial. Yet he boldly tweeted, "Guilty! He's guilty! I can tell!"[53]

When investors sued Stoam Holdings, the judge warned jurors not to do online research, but allowed them to use cell phones during the breaks.[54] A Wal-Mart employee serving as a juror, Jonathan Powell, used the breaks to tweet about the case. He set the stream so that he would be able to send out the tweets but not look at any incoming information. Most of the tweets were the inane sort that make up the bulk of the millions of tweets sent each day. But on the day the verdict came down, he tweeted the following: "'So, Jonathan, what did you do today?' Oh, nothing really. I just gave away TWELVE MILLION DOLLARS of somebody else's money!" Soon after, he posted another tweet: "Oh, and nobody buy Stoam. It's bad mojo, and they'll probably cease to exist, now that their wallet is $12M lighter."[55]

Powell says that "all Stoam-related tweets were made after the verdict had been handed down." But Stoam's attorneys are seeking a new trial based on his posts before, during and after the trial.[56]

But if a juror leaks information, how should that be dealt with? If someone tweets a verdict before it happens, should it be like insider trading, where a person can be jailed if he personally benefits or helps others benefit by inside knowledge of a company's plans?

People live their lives on social networks and it's hard for the judicial system to stop the knee-jerk reaction to friend whomever you want and say whatever you want. When Baltimore Mayor Sheila Dixon was tried for embezzlement, five of the jurors friended one another on Facebook. Maryland Circuit Court Judge Dennis Sweeney looked into the matter, questioning the jurors. A male juror a generation younger than the judge then posted on Facebook, "F—the Judge." When the 65-year-old judge questioned him further, the juror said, "Hey Judge, that's just Facebook stuff."[57] Dixon then entered a guilty plea, and the judge did not have to decide whether jurors' Facebook friendships are grounds for a mistrial.

With a generation gap around the use of social networks, new mechanisms need to be put into place to make sure that jurors refrain from seeking and disseminating information related to a case. Doron Weinberg, a lawyer for Phil Spector, wants jurors to have to provide the court with their online IDs—for Facebook and Twitter and other social networks—so that judges and lawyers can monitor whether the jurors are violating bans on blabbing about the case. But such a step seems a serious infringement of jurors' own free speech rights when they themselves aren't even on trial.

Other defense attorneys suggest keeping jurors off the internet entirely during the entire duration of the trial. At first blush, this seems similar to the already-established admonition to jurors not to read newspapers or watch television during the trial. But with so much of life from work to play occurring on social networks, asking people to give up the Web entirely might be like asking them to stop using the phone or to take a vow of silence. Such a requirement would violate our Social Network Constitution by infringing the right to connect.

Traditionally, the legal system has relied on the judge's instructions to the jury to convey the dos and don'ts of jury service. But jurors need more than just a standard instruction. They need to understand why inputs and outputs from outside the courtroom are forbidden. The librarian Gretchen Black said she knew they weren't supposed to look up facts about the parties but didn't realize they couldn't look up facts about the evidence. Juror Susan Dennis said she knew they were forbidden from tweeting but didn't know they couldn't blog. As a prospective juror, she'd blogged that the prosecutor was "Mr. Cheap Suit" and "annoying," while the defense attorney "just exudes friendly. I want to go to lunch with him. And he's cute." When Reuters brought her blog to the attention of the court, she was dismissed from the case.[58]

Courts are responding with new jury instructions telling jurors precisely what they should and shouldn't do—and why. The American College of Trial Lawyers emphasizes the importance of telling jurors why following the rules is important. It suggests that judges say, "The court recognizes that these rules and restrictions may affect activities that you would consider to be normal and harmless, and I assure you that I am very much aware that I am asking you to refrain from activities that may be very common and very important in your daily lives. However, the law requires these restrictions to ensure the parties have a fair trial based on the evidence that each party has had an opportunity to address. If one or more of you were to get additional information from an outside source, that information might be inaccurate or incomplete, or for some other reason not applicable to this case, and the parties would not have a chance to explain or contradict that information because they wouldn't know about it. That's why it is so important that you base your verdict only on information you receive in this courtroom."[59]

The American College of Trial Lawyers has also created a form that jurors should sign acknowledging their responsibility not to consult social networks. Its materials even include a model message that jurors can send friends and relatives telling them not to forward any information about the case nor ask for any comments until the case is over.[60]

As to output, the Administrative Office of the United States Courts has model jury instructions that specifically tell jurors not to discuss the case on Twitter, Facebook, Myspace, LinkedIn, and YouTube. The Ninth Circuit Court of Appeals, the federal appellate court that sits in the Bay Area, home to Silicon Valley and many web entrepreneurs, has similar rules for jurors but does not name the social networks to be avoided, assuming that new social networks will emerge that might replace or augment the current big five.

Knowledgeable court insiders wonder if the new instructions worsen the problem. The instructions might actually plant the idea of using social networks in jurors' minds. "It's like telling kids not to put beans up their nose," Abel Mattos, chief of the court administration policy staff at the Administrative Office of the United States Courts, told California's *Daily Business Review*.[61]

Judges and lawyers can lose their right to practice law when they use social networks in improper ways. But what penalties can be used against jurors who commit digital misconduct or upend trials by careless use of social networks? Banning them from being jurors isn't much of a punishment since many people don't want to serve on juries in the first place.

So far the punishments have been minor. When a Georgia juror Googled information in a rape case, the judge fined her $500. As with jurors' online research, judges are reluctant to imprison jurors who blab about a case. When a Michigan juror posted on Facebook, "Gonna be fun to tell the defendant they're GUILTY," the judge replaced her with an alternate and made her pay $250 and write a five-page essay about the Constitutional right to a fair trial.[62]

California, though, is getting tougher. Beginning in 2012, a new state law provides for a penalty of up to six months in jail for a juror who disobeys a judge's ban on the use of social networks, tweets, or web searches to find out about—or discuss—a case.[63]

Judges have generally been unwilling to put jurors in jail, even when they've cost the state hundreds of thousands of dollars in expense when a case needs to be retried. But if jury instructions are perfectly clear—and jurors are told why bans on social network input and output are important during trials, perhaps throwing a few jurors in jail might convince other jurors not to stray.

Jurors' misuse of social networks raises questions about whether jurors all along have disobeyed judges' orders and violated defendants' rights to a fair trial. Does the public nature of a tweet just alert us to a problem that was there before the Web was even imagined?

For generations, jurors have probably discussed cases with friends and family. Some may even have gone to a library or consulted an expert to help them make their decisions. But with the ease of access to inputs and outputs, the internet has made jurors' breaches easier and more likely. The dramatic cases that have arisen so far provide an opportunity to refresh our collective recollection about why the right to a fair trial is so important. In our Social Network Constitution, we should clarify that just as with any Constitution that guarantees rights, those rights come with responsibilities and obligations not to abuse them. We've already seen how rights end when an individual abuses them, such as when using speech to push someone towards suicide. As in the offline world, cyber rights should end at the moment when the exercising of those rights infringes on another's right. So, despite the right to connect, for the right to a fair trial to remain a reality, judges and lawyers should be disciplined for their misuse of social networks. And courts should create better instructions and be willing to penalize jurors for ignoring the rules.

12

The Right to a Fair Trial

When police responded to a burglary call in Martinsburg, West Virginia, the robber had done more than steal two diamond rings and ransack some cabinets. In the middle of the heist, he'd checked his Facebook page on the victim's computer and then forgotten to close the page.[1] The cops knew exactly who to arrest and did so.

But did they get the right guy? In this case, it seems they did. But think about how easy it would be to frame someone by accessing his or her Facebook page to show that a crime has been committed. Each of us probably knows or could guess the passwords of some of our friends or family members. Or we could gain access to someone else's email or social network pages by answering a few security questions based on publicly available information. A 20-year-old college econ major, David Kernell, was easily able to access Sarah Palin's email account by resetting her password using publicly available sources to answer routine security questions, such as her birthday, zip code, and where she met her husband.[2] He then posted her new password for anyone else to use.[3]

We all have a Fifth Amendment right not to incriminate ourselves. We can "plead the Fifth" and not testify. But can our Facebook and Myspace postings be used against us? And are courts savvy enough to know when someone else has posted something just to incriminate an innocent suspect—or to create a false alibi? How well does the justice system distinguish between the second self that we create on the Web due to wishful thinking (or the desire to appear tough or rich or young) and our real offline identities?

Anything you've already posted may come back to haunt you. Raymond Clark dressed as a devil for Halloween, his face painted red, black circles around his eyes, horns sticking out of his forehead. Countless other people across the country, including a psychiatrist friend of mine, similarly choose to wear a devil costume on that holiday. Many, like Raymond, posted those photos on their social network pages. But when Raymond became a suspect in the murder of Yale student Annie Le, the media's circulation of his Myspace devil photo made him appear guilty before he'd even gone to trial.[4]

Prosecutors have used Myspace and Facebook photos of people wearing gang colors or making gang signals to prove that they were involved in gang activity. But should the justice system really make that leap? A junior high student who was being bullied might post those kinds of photos to trick others into thinking he was tough. Under the Los Angeles Police Department's guidelines on gang colors, virtually all of us would fall into one gang or another. "Plaid shirts in either blue, brown, black or red" (worn by most of the college hipsters in the Ivy League) are indicative of gangs in LAPD-land. "Excessive amounts of dark clothing or a predominance of one-color outfits" (think of a New York art opening) "are also indicators of possible gang involvement."[5]

In Wisconsin, Jeremy Trusty's Myspace page contained a description of a short story about the murder of a judge.[6] That did not go over well with Judge Gerald Laabs, who was presiding over Trusty's divorce case. The judge arranged extra security for his courtroom. When a court employee heard Jeremy say to someone in the hallway, "We could just go in and shoot everybody," he was arrested.

At his trial for disorderly conduct, the prosecution introduced the short story description. Jeremy appealed, arguing that the First Amendment covered his Myspace posting, but the judge rejected the claim, saying that there was no precedent that protected such a posting. Yet doesn't the admission of the Myspace page in court get dangerously close to punishing the man for venting? And when we can be punished for our social network posts about what we are reading, where does it leave all of the millions of fans of thrillers and mysteries? (And where does that leave me, since I've written a novel in which a character blows up the White House?)

Social networks have become a cop's best friend—except, of course, when the cops embarrass themselves in their own postings. A survey by the International Association of Chiefs of Police of 728 law enforcement agencies from 48 states and the District of Columbia found that 62% of the agencies used social networks in criminal investigations.[7] Thieves have been identified when they've posted photos of themselves with stolen goods. Search requests, too, have helped to identify offenders. Nearly half of the law enforcement agencies said that social media had helped them solve crimes.[8] Robert Petrick's conviction for murdering his wife, for example, was secured through evidence from his Google searches, including "neck," "snap," "break," and a search for the topography and depth of the lake where his wife's body was found.[9]

Social network information has been used against people to show a crime was committed, to show that the defendant was prone to violence or lies, to impeach a witness, to show that the defendant or someone close to the defendant intimidated a witness, and to justify a longer sentence. But the current collection and use of social network postings in criminal cases runs afoul of the principles of both the U.S. Constitution and our emerging Social Network Constitution. The U.S. Constitution provides a wide range of rights regarding criminal justice. Some rights protect the public at large, and others protect the accused. But these protections against improper government action are routinely being ignored as social networks are used to uncover possible criminal behavior, to prosecute the offender, and to enhance the penalty.

The Fourth Amendment, which prohibits unreasonable searches of persons and property, was adopted to protect us from cops coming into our homes and searching for evidence of a crime by reading our private papers and going through our belongings. The Fourth Amendment also prohibits cops from routinely frisking us as we walk down the street. If law enforcement officials want to obtain evidence in that way, they need probable cause—some reasonable, individualized suspicion—that the particular person has committed a crime or is about to commit a crime. Most often, they also need a warrant, which they can obtain if they persuade the judge that their suspicions about the individual are reasonable.

With the Fourth Amendment, the Founding Fathers privileged privacy over crime prevention. Sure, the cops might miss out on finding an embezzler or blackmailer or wife beater by not being able to march into everybody's house just in case criminal activity was going on. But they'd probably have to unnecessarily invade the privacy of many upright citizens to find the people whose homes concealed evidence of a crime. And the Founding Fathers didn't want that to happen.

Cops can't even do such searches if they have evidence that, in the aggregate, people in a certain neighborhood or a certain group are more likely to be criminals. In some neighborhoods in a city, people are more likely to be carrying illegal concealed weapons. But that doesn't mean that cops can pat down everyone in the neighborhood to find weapons. The interference with innocent people's privacy is not worth it, under the Constitutional balance, even if the cops do find some criminals that way. A cop must have a reasonable suspicion about a particular person before he or she searches him or his home.

The Fourth Amendment also protects against discriminatory enforcement of laws. If cops weren't required to have individualized suspicion, cops might use stereotypes or prejudice to guide whose bodies or homes they search. A former editor of mine, at that time the highest ranking African-American man in publishing, often rented Jaguars when he traveled to other cities. He was invariably stopped by the cops, who thought that if a black man was driving a Jaguar, he must have stolen it. Without individualized suspicion, aggregate data can be applied in a discriminatory manner. Aggregate data shows that some drug runners at airports don't carry much luggage and look nervous. In one case where an African-American man was searched based on that criteria and actually found to have drugs on him, a dissenting judge railed about why the search was improper. He pointed out that he, too, is sometimes agitated when he flies but faces little chance of being stopped by the police. He remarked, "Perhaps it is my dress and manner; I believe it is these factors combined with the fact that I am white."[10] And think of how many innocent black men had been stopped to find one who was carrying drugs.

For those who think that targeting particular groups makes sense because some groups have a higher crime rate, the results of many studies prove the contrary. In Florida, for example, a study found that white women were slightly more likely to abuse drugs while pregnant than were black women. But black women were about ten times as likely to be prosecuted.[11] Another study showed that 80 to 90% of persons arrested for drug offenses are young black men, yet only 12% of drug users in the U.S. are in fact black.[12]

Even though offline searches can't be conducted without probable cause, cops routinely search social networks. Some criminal charges have been based entirely on photos found on social networks. A deputy sheriff who was also an employee of Glenbard South High School in the Chicago suburbs searched the Facebook pages of the high school athletes. When the deputy found party photos indicating underage drinking, four boys were suspended from their team and criminally prosecuted. The aunt of one of the athletes, at whose home the party was held, was charged with contributing to the delinquency of minors.[13] In another case, a 16-year-old boy posted a Myspace photo of himself with his father's guns. Even though it is legal to have a gun at home and his father had permitted him access to the guns, a jury convicted the boy of possession of a gun by a juvenile.[14]

Sometimes the digital dragnet is conducted not by the police but by private citizens. When a 19-year-old Florida woman, Rachel Stieringer, posted what she thought was a humorous photo of her 11-month-old baby holding a bong on Facebook, a Texas man contacted the Florida Department of Children and Families, which launched an investigation into her parenting skills. The baby tested negative for drugs, so he was not taken from the mom.[15] Yet the cops still threw her in jail for possessing drug paraphernalia.[16] But bongs are sold legally across the U.S. and over the Web by American companies like Grasscity.com. Should private citizens be able to launch criminal investigations based on Facebook photos? And, in the bong case, once it was determined that the woman's baby was safe, shouldn't the case have been dropped?

Sometimes cause-oriented individuals search the Web to find evidence of problematic behavior—such as abuse of animals. When members of the social network 4chan saw a photo that appeared to show a boy mistreating a cat, they used their computer skills to track down the identity of the boy and gave that information to the police.

Other animal rights activists traverse the Web for potential breaches of the law protecting endangered species. When two Americans, 24-year-old Alexander Rust and 23-year-old Vanessa Starr Palm, were vacationing in the Bahamas, they posted Facebook photos of themselves capturing an iguana and grilling some meat.[17] The pictures were brought to the attention of the Bahamian police, who arrested the Americans for killing an endangered species. Iguanas are protected not only under Bahamian law but under the Convention on International Trade in Endangered Species of Wild Fauna and Flora. As a result, according to Eric Carey, the executive director of the Bahamas National Trust, the two Americans "could also be charged under U.S. law, which makes it illegal to commit an offense in a country that has a relationship with the U.S."[18]

Should the two Americans be put in the slammer based on the photos alone? Not all iguanas are endangered and it's unclear whether they were flagrantly breaking the law or just trying to reenact an episode of *Survivor*. A recent peek at Vanessa's Facebook page reveals that she is now part of the group Stop Animal Cruelty. But does that posting indicate remorse—or is it just an attempt to appear less culpable?

In other instances, law enforcement agencies create "virtual deputies" to use social networks to identify lawbreakers. Such was the case in Texas, where cameras were installed along the border and people could log on to watch and report illegal border crossings. The cameras covered public spaces. But what if a virtual deputy started collecting evidence in places the police couldn't, like by pointing a camera at a neighbor's window?

Letting ordinary people become virtual deputies patrolling social networks brings us dangerously close to the police state portrayed in the film *The Lives of Others*, which won the Academy Award for Best Foreign Film in 2007. In the film, set in Germany in the mid-1980s, the secret police, the Stasi, have bugged a playwright's apartment. The agent listening in uncovers evidence of a crime—the playwright had an unregistered typewriter on which he'd written an article critical of the government. But the small amount of incriminating information the agent hears is dwarfed by the sheer number of private, intimate details he learns about the life of a married couple.

In Germany in the 1980s, as in the 1950s McCarthy era in the United States, neighbors informed on neighbors. After the fall of the Berlin Wall in 1989, Friedrich Hans Ulrich Mühe, who played the Stasi agent in *The Lives of Others*, found his own Stasi file, which indicated that fellow actors had informed on him and that his wife had been tricked into giving information about him.[19] When Mühe was asked how he later prepared for his role in *The Lives of Others*, he said, "I remembered."[20]

When citizens scour social networks for incriminating evidence, their actions raise many of the same issues as when searches are done by cops. The privacy of many innocent people is invaded. People who did not commit a crime but whose photos or posts are referred to the police face the stigma of being investigated. And most important, these citizen searches are likely to be undertaken in racially discriminatory or vindictive ways.

In many instances, citizens send social network information to cops to be vindictive or to gain something themselves. About half of teenagers post photos of themselves engaging in illegal behavior, mainly underage drinking. Yet not all those teens are prosecuted. Who passes on those incriminating photos to the cops? Often it is the parent of a rival to the teen in the photo. If the photographed teen is suspended from school or knocked off the team, the rival will gain a spot. When Ashley Payne was pressured to quit her job because of a photo of her drinking a Guinness on her vacation visit to the beer company's Irish factory, it was likely a jealous fellow teacher who anonymously sent the photo to the superintendent of schools.

What standards should govern cops' access to people's Facebook pages? Our Social Network Constitution deems social networks to be private spaces, so law enforcement officials would need to have probable cause before they searched social networks. And law enforcement wouldn't be allowed to get around those rules by "deputizing" citizens to conduct searches or find offenders themselves. Protecting against cyberdragnets—either by law enforcement or by private citizens—is in keeping with the high premium that our Social Network Constitution puts on privacy.

Once social network evidence gets to court, there are further issues. The Sixth Amendment of the U.S. Constitution provides for the right to a fair trial, and the Fifth Amendment provides that no person shall be "compelled in any criminal case to be a witness against himself, nor be deprived of life, liberty, or property, without due process of law." Those Constitutional rights help govern what information is admitted in criminal cases. Judges are supposed to make sure that the evidence is authentic, relevant, and more probative than prejudicial. But as in family law cases, evidence is being admitted in criminal cases without meeting the basic evidentiary requirements.

Let's say I repeatedly post, "I hate my husband and some day he'll get his." Consider four different scenarios. A few months later, I divorce him. A few months later, he and I get into a horrible fight, he moves to strike me, and I reach into the bedroom drawer, pull out a gun, and shoot him in the heat of the moment. A few months later, I shoot him after planning how to go about it for weeks. Or, a few months later, he shoots himself, committing suicide.

In the last three scenarios, where he ends up dead, what weight should be given to my statement of hatred? It is likely that it will be used to say I acted with premeditation (which could mean the death penalty for me in many states) even if I shot him in the heat of the moment—or even if he shot himself. Looking at social network posts in retrospect might not produce reliable evidence.

Yet social network posts and other digital missives have already been used to provide evidence of a crime or show premeditation. Maurice Greer found himself on the receiving end of an aggravated murder charge due in part to a previous Myspace status posting that said, "Pow, one to the head, now you're dead," accompanied by a photo of himself with a gun in the waistband of his pants.[21] A cop testified that the gun looked like the same one that was used to kill a victim. But think about how radically different a measure of proof that is than what is generally used in forensics. Generally, a gun would be linked to a defendant only through extensive scientific testing such as fingerprinting and ballistics. With a social network, a connection was made between a murder weapon and an indistinctive handle of a gun sticking out of his pants. In that case, there was plenty of other evidence to link Maurice to the crime, but the appellate court's nod of approval to the use of the post and the photo sets a dangerous precedent that could lead to innocent people being convicted based on seemingly incriminating evidence about their digital selves.

At sentencing, social network information again comes into play. In a Rhode Island case, a 20-year-old's sentence was harsher because of pictures posted to Facebook. The pictures were taken two weeks after the defendant, Joshua Lipton, drove under the influence of alcohol and caused a car accident involving three cars that seriously injured a woman.[22] They portrayed him wearing a Halloween costume that consisted of a bright orange jumpsuit with the phrase "jail bird" written in black, with a black-and-white-striped undershirt.[23] One of the victims of the accident saw the photos and gave them to the prosecutor. At the sentencing hearing, the prosecution argued that the pictures demonstrated that Lipton was not remorseful and thus deserved incarceration, as opposed to probation, and the judge agreed.[24]

Some courts use social network information—such as photos indicating potential gang affiliation—to enhance sentences. In Arkansas, a photo of Brittany Williamson making gang signs was used to enhance her sentence for battery when she and three friends beat up a 14-year-old teen to steal his necklace.[25] In contrast, a California appellate court held that the trial court judge should not have allowed the prosecution to introduce a photo of a witness in gang attire and an alleged social network roster of members that had purportedly been posted by the gang itself.[26] The appellate court reduced the defendant's sentence to eliminate the extra years that were tacked on due to gang membership. The court said that social network photos need to be authenticated—for example, by someone who was present at the time the photo was taken or by someone who could state that the photo had not been altered.

As with custody cases, where evidence from social networks about a woman's sexuality has improperly been used against her, postings and photos that show a female victim's sexuality have been used to undercut her testimony. Even though evidence of prior sexual activities or the clothes the woman wears is generally not admissible, accused rapists are now getting around that rule by introducing sexy statements and photos of victims from Facebook and Myspace. They argue that the postings are being used not to show past sexual exploits but rather to question the victim's credibility.

For young girls who are victims of sexual abuse by a male relative or their mother's boyfriend, a Myspace posting of a sexy photo can do damage to the jury's opinion of her, suggesting that she asked for it. But should such a photo even be admitted into evidence? It's not uncommon for young girls to become hypersexualized after being sexually abused. After all, one of the psychological impacts of incest is to make the girl feel like her only value is as a sex object.

Social networks seek out all sorts of relationship information from users. Numerous surveys ask participants in Myspace and Facebook about their sexual experience. A teenage girl or boy might not answer those questions honestly and may instead decide to claim sexual experience in order to appear more sophisticated.

Sometimes a teenager's friend fills out the survey for him or her. (Even some adults allow their friends to create a Myspace profile for them.)

Should photos and sex surveys be admitted to impeach an incest victim? When a 13-year-old girl finally got up the courage to tell a teacher that that she'd been repeatedly raped by her father, a physical exam showed that she'd been forcibly sexually abused.[27] She and her brother also showed evidence of having been beaten by her dad with a belt. The father would have her strip naked before the beatings, and then, about a year before she spoke to the teacher, he started forcing sex on her. Each time, he'd say it was going to be the last time, but his sexual abuse continued. When the father was convicted, he appealed, claiming that the trial court made a mistake by not allowing him to introduce evidence from the girl's Myspace page. He wanted a new trial so that he could introduce suggestive photos of the girl from her page. He also wanted to introduce the results of a survey where she answered in the affirmative to the question, "Had sex?" The girl said that the survey had been filled out by one of her friends.

The appellate court held that despite the rule that prior sexual activities should not be admitted in rape cases, the Myspace page should have been admitted as evidence to address the girl's credibility. The court didn't grant a new trial, though, since there was ample other evidence of the father's sexual abuse. But its pronouncement that someone's sexy photos and survey answers from a social network should be admitted in sexual abuse cases sets a troubling precedent for women's ability to get justice when they are raped.

In contrast to rape cases, some courts will keep out information from social networks or websites if the information is likely to be prejudicial without sufficient relevance to the case.[28] For example, an African-American defendant charged with mail fraud tried to argue that in fact a white supremacist group committed the act. She pointed to racist rantings on the Web that were supposedly linked to the groups. The judge held that the defendant could not prove that the groups had actually posted the remarks and that the remarks would be prejudicial.

What might be some valid uses of social network posts and photos by the legal system? When a woman was already wanted on an arrest warrant but had fled the jurisdiction, cops monitored her high school reunion page and caught her when she came back from the reunion. When there was evidence that two foster children were being abused, further evidence was collected from a Myspace page. The court allowed the admission of evidence from the foster father's Myspace chats with the children, saying such things as "the bear gets the twinkie at midnight," which the prosecution alleged was a reference to sex with the boys.[29] The court held that the chats could be authenticated through the testimony of one of the victims.

Social network posts are also relevant and should be admitted—if they can properly be authenticated—when the crime itself is one that occurs in the cyber realm, such as the William Melchert-Dinkel case of pressuring people to commit suicide. Social network posts and other electronic communications have also properly been used to show improper attempts to influence a juror or intimidate a witness.

When Miguel Lopez went on trial for grabbing his girlfriend, Angela Gonzales, by the hair and battering her head against the floor of their apartment while she held their 16-month old child,[30] Angela tried to get Miguel off by tracking down a juror through Myspace and pleading with the juror to find Miguel not guilty of assault.[31] Initially, Angela Gonzales told police that Miguel became enraged when she informed him that she did not have enough money to buy cigarettes.[32] Later she tried to say that he had never struck her and that the swelling and bruising on her face resulted from vomiting due to migraine headaches[33]—despite the fact that he'd assaulted her several times in the past.[34] After the juror told the court about the Myspace communication, the court excused the juror from the case and the jury convicted Miguel Lopez of felony domestic violence.[35] Angela Gonzales ended up in jail as well—for jury tampering. District Attorney Jeff Reisig said, "It is regrettable that a victim of a crime must be charged with a crime herself, but communicating with an empanelled jury risks the integrity of the justice system and cannot be allowed."[36]

When Antoine Levar Griffin, aka Boozy, was retried for murder for shooting another man in the bathroom of a Maryland bar, Antoine's cousin Dennis Gibbs testified that he'd seen Antoine with a gun, chasing the victim into the women's bathroom.[37] Antoine's attorney challenged the witness's credibility since he'd told police at the scene—and told the jury in the initial trial—that Antoine had not gone into the bathroom.

That's when Dennis explained that he'd lied earlier because of fear. Antoine's friends had threatened him and, he said, "then right before the last [trial] I was threatened by Antoine's girlfriend. She told me I might catch a bullet if I showed up in court."

When the defense attorney questioned whether the threat had actually occurred, the prosecutor offered a Myspace profile he said belonged to the girlfriend. The profile page warned: "FREE BOOZY!!!! JUST REMEMBER SNITCHES GET STITCHES!! U KNOW WHO YOU ARE!!"

The defense attorney pounced on the evidence. How did they know that the page was really the girlfriend's? It was just listed as belonging to someone called SISTASOULJAH. And they only had the word of a cop who printed it off the Web that this was even a real Myspace profile.

Outside of the presence of the jury, the judge allowed the defense attorney to question the cop.

"How do you know that this is her web page?"

"Through the photograph of her and Boozy on the front, through the reference to Boozy [the nickname of the defendant], to the reference of the children, and to her birth date indicated on the form."

The judge let in the Myspace page, over the defense attorney's objections. And when Griffin was convicted of second-degree murder, his attorney brought an appeal. In May 2010, the appellate court noted that there were no Maryland prec-edents about how to judge the authenticity of social network evidence and "scant case law from other jurisdictions." The court said, "We have not found a reported appellate decision addressing the authentication of a printout from a Myspace or Facebook profile."[38]

The court said it could see how social network information could be "invalu-able" since "the design and purpose of social media sites make them especially fertile ground for 'statements involving observations of events surrounding us, state-ments regarding how we feel, our plans and motives, and our feelings (emotional and physical).'"[39]

The appellate court also saw some difficulties. Unlike the early days of the Internet, when AOL posts could be traced to a person through billing records, anyone can post on today's Web without having to reveal his or her identity. But here the photos and facts seemed to match up to the defendant's girlfriend. And the threatened witness himself could corroborate that he'd seen the post on Myspace. The court was willing to rule that social network posts could be "circumstantially authenticated" based on "content and context."

The court waved aside defense concerns that the social network posts could have been altered. But there have been cases where one side planted fake social network or email evidence against the other side. Danielle Marie Heeter created a fake Myspace account in the name of her husband's ex-wife and sent scathing mes-sages to herself,[40] hoping that it would cause the ex-wife to look bad in her custody dispute over her son, Ryan. The messages called the new wife a "cunt" and said, "Oh, and another thing, when I get custody of my son, I will make sure you are not allowed around him. Like I told you before, it's—before, it's just better if you just disappear."[41]

When confronted by police, the new wife denied that she'd sent the messages. But prosecutors showed that the messages came from her computer at a time when she was logged on to her own Myspace account. A jury convicted her of felony fraud (for which she was given probation) for attempting to influence the custody case by making it appear that the ex-wife was unstable and ruthlessly cruel.

What if the court had believed that the posts were made by the ex-wife? She could have lost custody of her son if she wasn't able to launch an investigation into the location where the posts had originated. And not everyone's going to be able to afford to hire an online data expert to assist at trial. During the "social media trial of the century," a defense expert's analysis of some 40,000 blogs, social network posts, and tweets helped Casey Anthony's lawyer tailor his trial strategy.[42] The prosecution did not have a social media analyst, but the trial consultant for the defense team, Amy Singer, monitored online posts to gauge public opinion about everything from what color Casey looked best in to why Casey was guilty. When bloggers began criticizing George Anthony, Casey's father, the defense team began questioning him more aggressively. According to Singer, the online posts gave the defense team a sense of how much it could "blame George." Similarly, when online sentiment turned against Cindy Anthony, Casey's mother, after she testified that she had performed the chloroform searches attributed to Casey, the defense team "distanced" itself from her. According to Singer, who provided trial consulting services for the O. J. Simpson trial and Jack Kevorkian case, "Social media was the difference between winning and losing."[43]

In the Anthony trial, the prosecution introduced evidence that the Anthony's family computer had been used on 84 occasions to visit a website about chloroform.[44] The defense attorney did not call upon any expert to dispute that evidence.[45] But he should have. When the prosecution expert reanalyzed his own data before the trial ended, he found that there had been only one search.[46] But, says the expert, when he contacted the prosecutors, they made no effort to correct the record.

The social network issues don't go away once the person is convicted. Sometimes whether and how the offender can continue to use social networks becomes part of the sentence. When William Melchert-Dinkel's instant messages were used to convict him of pressuring Nadia Kajouji to commit suicide, the judge forbade him from future use of a computer without court permission. Other judges require convicted offenders to friend them. Judge Kathryn Lanan of the Galveston juvenile court requires juveniles under her jurisdiction to friend her and calls the kids back to court if she discovers inappropriate conduct.[47] And a Michigan judge, A. T. Frank, searches social network sites of offenders on probation; he has found Facebook and Myspace photos of violations, including drug use.[48]

Criminal defense attorney Dennis Riordan finds such judicial action troubling. "Sure, probation officers have a right to visit an offender's home. But the point is that they take their findings to a judge, who then decides if what they found warrants revoking probation and putting someone in jail. If a judge friends an offender, that judge is acting as both the investigator and the court. It's troubling."

Social networks increasingly play a role in criminal cases, without adequate consideration of the consequences. Judges treat social network posts as gold-plated evidence, when they could actually be faked by the real offender or created for a different purpose, like the hateful emails of the new wife made to look like they were coming from the old wife or the Myspace page of the school principal created by a student that made him seem as though he were a pedophile. Our Social Network Constitution should apply the right to privacy to forbid social network dragnets unless there is probable cause to think a crime has been committed. When social networks enter the courtroom, judges should apply the right to a fair trial and should consider how the information was obtained, whether it was falsified or authentic, and how closely related it is to the issues in the case. Unless the social network posting provides reliable evidence of some element of the crime (not just some prejudicial information about other aspects of a defendant's or a victim's life), that posting should not be admitted.

13

The Right to Due Process

"Stop spreading lies about Iran on Facebook," one message addressed to Hamid, an Iranian-American graduate student, stated in Farsi. "We know your home address in Los Angeles. Watch out, we will come after you." The email, signed by "Spider," was sent after Hamid changed his Facebook profile picture to an image that showed his solidarity with the Iranian opposition. When Hamid posted that image—a Victory sign trickling blood—he thought it was available only to the people closest to him, the people he'd friended. But Facebook had changed its privacy settings to make certain types of personal data, including a person's list of friends and profile pictures, available to the public.[1] This sudden revelation of users' personal information had dramatic repercussions.

Dozens of American Facebook users who had posted Facebook messages crit ical of Iran reported that the Iranian authorities questioned and detained their relatives.[2] After Koosha, a 29-year-old engineering student, attended rallies in the United States and created an online petition for the release of a human rights lawyer detained in Iran, he began receiving threats that his relatives in Tehran would be hurt. Then, Koosha's mother called, saying that his father had been arrested by security agents in Tehran.[3] His father has now been released, and Koosha says he takes care not to talk politics when he contacts his parents.

In the United States, too, users of social networks faced troubling repercussions after their personal information was made public. In 2010, Google sought to create an alternative social network to Facebook. Its email system, Gmail, already had 176 million users, so Google had a built-in audience, and each Gmail user had a circle of friends—the people they emailed and chatted with most frequently.[4] When Google Buzz launched, Gmail users got a message about the new service and were given two options before they were able to access their email inbox: "Sweet! Check out Buzz," and "Nah, go to my inbox."[5] Users who selected "Sweet! Check out Buzz" were shown a welcome screen and told that "You're set up to follow the people you email and chat with the most."[6]

Without giving adequate notice to users, Google Buzz made users' personal information available to the people on its automated list—basically the people you had the most Gmail contact with.[7] The automated list of your top Gmail contacts could be viewed by others on the list,[8] providing a potential glimpse into your intimate relationships. That could include the person you have a restraining order against, your psychiatric patient, a lawyer you'd contacted in contemplation of divorce, or the job recruiter who emailed you a lead.[9] Those frequent email contacts also could see what you were posting on the photo site Picasa and what you were posting on the Google blog site Reader.

As a result of Google Buzz, the abusive ex-husband of a female blogger was able to find out her current location and workplace and gain access to communications between her and her new boyfriend.[10] A business owner was shocked to see that the identities of his most frequent contacts—his clients—were made public, thus giving vital information to his competitors.

Without adequate notice, even the publication of seemingly benign information can disrupt people's lives. A 2007 Facebook project called Beacon broadcast information on the Facebook news feed about a user's purchases at affiliated websites, such as eBay, Overstock, Hotwire, and Blockbuster.[11] Facebook did not give users adequate warning that detailed descriptions of their purchases would be publicized to all of their Facebook friends.[12] When a man bought a diamond ring online, that fact was broadcast to his 220 Facebook friends and 500 classmates: "Sean Lane bought 14k White Gold 1/5 ct Diamond Eternity Flower Ring from overstock.com."[13] His wife queried, "Who is this ring for?" The ring was for his wife, but her surprise Christmas gift had been ruined.[14] Facebook shut down Beacon amid public outrage[15] and settled a lawsuit brought by plaintiffs like Sean, who alleged that Facebook had failed to "provide notice and obtain informed consent" before posting their personal information.[16]

Social networks routinely change the rules midgame, cheating unsuspecting users out of an opportunity to control their information and ultimately their fate. From Iran to the United States, the publication of information without warning and consent has jeopardized people's freedom, safety, and relationships.

Our Social Network Constitution secures a bevy of rights—from the right to connect to the right to privacy. But substantive rights like free speech, privacy, and anonymity can be demolished when proper procedures aren't put in place. Policies are needed to allow people to make choices about their information and control who has access to their second self. Under the U.S. Constitution, for example, the Fifth and Fourteenth Amendments give people the right to due process of law. This requires adequate notice of the rules, proper procedures before rights are denied to you, and enhanced protections to prevent certain rights from being denied at all. Similarly, our Social Network Constitution needs not just substantive rights but procedures to protect those rights.

The first crucial procedural right is one of notice. Everyone knows about the Miranda warning in criminal cases. Unlike the Facebook privacy controls, it's not buried 170 steps away or as part of its extensive and complicated terms of service.[17] It's not a 45,000-word document, like Facebook's privacy policies. It includes the simple statement, "You have the right to remain silent. Everything you say can and will be held against you in a court of law." We need a Miranda-type privacy warning for social networks. Something along the lines of "You have the right to remain private," with an explanation of how social network information might be used against you in each instance that you expand your audience. That way, people will have an idea of what they might be giving up or getting into if they agree to wider dissemination of their information.

Right now, people do not receive adequate warning about the implications of posting on social networks or using the Web more generally. Sure, there are those long notices that we have to click before we enter a website, but most people don't read them or don't understand them. Gamestation, a gaming company, put the following clause into its terms and conditions as a joke:

> By placing an order via this Web site on the first day of the fourth month of the year 2010 Anno Domini, you agree to grant Us a non transferable option to claim, for now and forever more, your immortal soul. Should We wish to exercise this option, you agree to surrender your immortal soul, and any claim you may have on it, within 5 (five) working days of receiving written notification from gamestation.co.uk or one of its duly authorized minions.[18]

The overwhelming majority of people (88%) agreed to those conditions, leading Gamestation to assume that people don't read terms and conditions. Or perhaps it's that data aggregators and social networks already have your "immortal soul," as you're captive in an increasingly digitized world of surreptitious collection of data without notice, consent, or an opportunity to be heard.

In the realm of data aggregation, you generally don't get any notice that data about you are being collected. When your internet service provider contracts with a data aggregator so that the aggregator is able to intercept and use every single bit of information you send, you are not told of the tap, nor are you asked for your consent. Commercial interceptors have even more powers than the police, who would need to get a warrant from a judge before they tapped your lines.

Under the U.S. Constitution, people have the right to advance notice of government policies that will affect them. To comply with the U.S. Constitution, when legislators adopt a law, it needs to clearly express what behavior or actions are allowed or prohibited. According to the U.S. Supreme Court, "a statute which either forbids or requires the doing of an act in terms so vague that men of common intelligence must necessarily guess at its meaning and differ as to its application violates the first essential of due process of law."[19] The ability of individuals to understand policies in advance, before those policies can be applied to them, is key.

The right of advance notice is also present in the private realm. Banks need to provide you with information before you agree to a loan. Before a doctor can undertake treatment on you, she must provide you with sufficient information about your disease, the risks and benefits of treatment, and any alternative to the treatment. For example, a patient in need of an operation must be briefed before the surgeon cuts him open. An individual with high cholesterol who is participating in a drug trial must understand the risks of taking either the experimental drug or the placebo.

The principle of advance notice is virtually nonexistent on social networks. In 2006, Facebook, without informing its users, began broadcasting its users' personal information (including details about users' marital and dating status)[20] through its "News Feed" program.[21] Hundreds of thousands of Facebook users objected to the change.[22] Mark Zuckerberg sent an open letter in response, "We really messed this one up. When we launched News Feed and Mini-Feed we were trying to provide you with a stream of information about your social world. Instead, we did a bad job of explaining what the new features were and even worse job of giving you control over them."[23] Doing a "bad job" of explaining new features that disclose personal information and failing to give users an easy way to control the disclosure of their personal information has become the modus operandi for Facebook.

Some courts have started to apply the requirement of notice in the digital realm. A consumer sued the computer services company Netscape for invasion of the user's privacy by using a software plug-in program that was supposed to facilitate internet use and was available for free download on Netscape's website. Netscape said that the consumer was bound by the terms of service. Those terms were hidden in a footer on the website and did not require any affirmative act of the consumer. The court rejected Netscape's argument, saying, "There is no reason to assume that viewers will scroll down to subsequent screens simply because screens are there. When products are 'free' and users are invited to download them in the absence of reasonably conspicuous notice that they are about to bind themselves to contract terms, the transactional circumstances cannot be fully analogized to those in the paper world of arm's-length bargaining."[24] The court held that if a link to a license is not conspicuous enough on a web page, then even if a user clicks on the link, the website can't assume mutual assent to the contract.

Not only should the notice be conspicuous, it also should be simple and easy to understand. Consider Gamestation's terms of service, where people didn't even notice that they were giving away their immortal souls. Or think about Facebook's 45,000-word privacy document. Showing clear warning labels is not a new concept. For years, cigarette companies have been required to display a Surgeon General's Warning, which aims to educate smokers about the risks of smoking. In September 2012, the FDA will require the warnings to be accompanied by graphic images that more clearly illustrate the consequences of smoking.[25]

The second vital procedural right is the ability to control your personal information and second self. In the medical context, not only must a patient receive adequate notice before treatment, he must be allowed to make a voluntary, noncoerced decision of whether or not to participate.

But on social networks and elsewhere on the internet, cookies, web beacons, deep packet inspections, and other means employed to surreptitiously collect your data occur without warning and without your consent. Almost every time you visit a website, at least one tracking mechanism is installed (and on some websites, such as Dictionary.com, hundreds are installed). Not only are those mechanisms able to watch what you do, they take your personal information, store it, and transmit the data to a company that either uses or sells it.

As a user, you should have the ability to control what happens to your information. Right now, saying no to web tracking is difficult, if not impossible. First, you may not know that information about you is being collected. Second, you may have signed on to a social network under one set of rules and, without warning, those rules may have changed. Third, saying no may require hundreds of steps. And if you try to get rid of tracking mechanisms from data aggregators by setting your browser to prompt you every time a third party puts a cookie on your computer, you will spend all day fending off cookies—and you may be extorted to accept them by websites and social networks that will allow you access only if you agree to share information with unknown third parties. The principle of control would allow the users of social networks to prevent third parties, or even social networks themselves, from collecting or using user information.

Your right to control is currently being diluted by the sheer number of things you have to do to exercise it. The current system is based on the idea of opting out. But according to a 2008 Consumers Union report, most Americans don't agree with that type of policy. A Consumer Reports poll revealed that 93% of Americans think that "internet companies should always ask for permission before using personal information," conveying the idea that social network users should have more control over their information and should be given an opportunity to opt in before any data is collected.[26]

You should have to affirmatively go to a separate website to choose which third-party cookies or other tracking technologies you will allow to be used on your computer and with your data. Think of it like inviting people to your house. You shouldn't come home, find that a bunch of people have broken in, and have to ask them, one by one, to leave. Rather, you should be able to reach out and contact third parties if you want them to visit you.

Currently, even when you do think that you've finally opted out, you might be wrong. The Federal Trade Commission stepped in on the side of due process when the Chitika advertising network misled users about its collection of their personal information. Chitika, which means "snap of the fingers" in Telugu, a South Indian language, places more than 3 billion monthly advertisements on over 100,000 websites.[27] Chitika had this privacy policy:

> When users visit a page in the Chitika network, one or more cookies—a small file containing a string of characters—are set to the computer that uniquely identifies the users [sic] browser. Chitika uses cookies to improve the quality of the targeting

service by storing anonymous activity data and tracking user trends, such as how people search and browse. Users can reset their browsers to refuse all cookies or to indicate when a cookie is being sent. . . . Chitika encourages and promotes business practices that protect and honor the privacy of users. You can opt-out of receiving Chitika cookies by using the button below.[28]

But when people clicked to opt out, their wishes were followed for only ten days. Then Chitika would begin tracking the users with cookies, without notifying the users, whenever they visited any of the 100,000 websites in Chitika's advertising network. The FTC ordered Chitika to destroy any data previously collected through its cookies and to provide an opt-out mechanism that lasts a minimum of five years.[29]

In addition to data aggregators, social networks also play a role in collecting and sharing private information without consent. Facebook has often changed its stance on what types of personal information about each of us are made available to the public and third parties, often without any notice to its hundreds of millions of users. "I have listened to Facebook experts discuss the privacy settings who quickly became confused," Marc Rotenberg, Executive Director of Electronic Privacy Information Center, said in a presentation to Congress. "I even heard Facebook founder Mark Zuckerberg describe the new changes to his company's privacy settings only to learn, unexpectedly, that some of his college photos were now available to 'everyone.' I am convinced that not even Facebook understands how its own privacy settings operate."[30]

Due process is not only about notice or control of information and privacy. It's also about correcting or deleting any information that's out there. Under the U.S. Constitution, if the government makes a decision that disadvantages you (such as by taking away benefits because it no longer thinks that you are disabled), you initially have the opportunity to attempt to "correct" the government's conclusion and are entitled to a hearing to resolve the matter.[31] In the social network world, there is no guaranteed mechanism to correct or delete data about yourself unless you contract with a third-party reputation service, whose effectiveness isn't guaranteed.

On Spokeo, for example, the information listed about an individual might not be correct. Yet banks, credit card companies, employers, and other institutions routinely use weblining to determine whether or not to offer a job or extend credit based on such information. After Spokeo misrepresented Thomas Robins's age, marital status, profession, and other information,[32] he sued Spokeo for violating the Fair Credit Reporting Act.[33] The act requires that consumer reporting agencies provide a mechanism for individuals to request information about their file, dispute its accuracy, and correct inaccurate, incomplete, or misleading information.[34] A federal court is currently considering the case.

An individual should also have a right to delete his or her data. Under the U.S. Constitution, people have the right, in various situations, to remake themselves, such as by declaring bankruptcy. Almost all states have laws that allow juvenile records to be sealed, expunged, or deleted.[35] Some states also offer qualifying adults an opportunity to expunge or seal their criminal records.[36]

In the social network realm, deletion could be accomplished through a combination of direct rights (for example, letting an identified private individual delete photos of herself on the Web) and planned obsolescence. In his book *Delete: The Virtue of Forgetting in the Digital Age*, Viktor Mayer-Schönberger makes the provocative suggestion that we code in an expiration date to the photos and information we post on the Web.[37] If the picture of you chugging beer with your now ex-girlfriend evaporated after two years, think how much better your life might be.

In 2011, Representative William Straus introduced bill number 02705 to the Massachusetts House of Representatives.[38] Under that bill, personally identifiable information cannot be collected about you by a data aggregator unless you specifically opt in, and the maximum amount of time a third-party advertising network may retain even non-personally identifiable information is 24 months.[39]

In other countries people have a right to advance notice, to control, and to correct information. The European Union already has a directive that requires that entities not collect more data than they need for a particular transaction, that they ensure that the data is accurate and complete, and that identifiable databases are kept "no longer than is necessary for the purposes for which the data were collected."[40] And European Union Commissioner Viviane Reding is preparing to present a proposal to allow people to erase information about themselves on the Web. "Everyone should have the right to be forgotten," she said. "Due to their painful history in the 20th century, Europeans are naturally more sensitive to the collection and use of their data by public authorities."[41]

Advance notice, control, and the ability to correct and delete are imperative to ensuring that social network users are able to realize their substantive rights. Each policy change by social networks in the past has had serious consequences to social network users, such as the Iranian-American detainee or the woman trying to stay clear of her abusive ex-husband. While there have been clear violations of due process rights in the past, there is an even greater threat in the future as technologies push forward unregulated.

Social networks amass and control large troves of personal data. Soon Facebook alone will have data on a billion people. "Imagine having access to the political views, sexual preferences, relationships, tastes, foibles, emotional states, workplace attitudes, etc. of a billion people," writes Richard Power in the journal *CSO: Security and Risk*. "An effort to collect such data on behalf of a government, or a corporation, or a geopolitical alliance, or an industrial sector, or even a seemingly benign world organization, would meet with fierce opposition. It would be difficult if not impossible; it would require lawyers, money and yes maybe even guns."[42]

But we've all been herded peacefully into Facebook Nation without adequate consideration of our right of privacy or even our rights of property. In every other setting, we'd have a commercial property right over photos we took or missives we wrote, but in Facebook Nation, the company is commercializing our postings.

Facebook's policy changes often violate basic principles of due process. Even though, under the Nuremberg Code, before research is undertaken, "The voluntary consent of the human subject is absolutely essential,"[43] Facebook undertook biometric research on its users without their consent. Biometric data is information about an individual's unique characteristics; it allows identities to be confirmed based on that person's traits (for example, fingerprints or eye scans). The biometric data collected through Facebook's research categorized aspects of a person's face such as the distance between a person's eyes and other measurable aspects of a person's facial features and facial structure.[44] The research applied biometric algorithms to both public and private Facebook photos to develop its facial recognition program.

When Facebook introduced its facial recognition software to users worldwide, it silently enrolled all users in the program without their permission.[45] Now when a user uploads a photo, Facebook's software scans the photo and if it recognizes a friend's face it will suggest a tag that adds the user's friend's name.[46] Users can employ a cumbersome series of steps including three screens and seven clicks to opt out of having the software suggest that they be tagged in their friends' photos, but the default setting is for the facial recognition software to be enabled.[47]

There is no option in Facebook's privacy preferences to prevent the collection of biometric data.[48] Facebook still keeps the "summary information drawn from comparing any tagged photos of you." There is no way under "privacy settings" to completely remove the data Facebook has collected about you through facial recognition. To have your biometric data deleted, you have to contact Facebook through a difficult-to-find link on the site. To access this link, you need to go to Facebook's Help Center, type "photo tagging" into the search field and run the search, click on the "Information about photo tagging" link under Page Results, and expand the third link "How can I remove the summary information stored about me for tag suggests?" A contact link is available in the first sentence of the answer. You must click on the "contact us" link and send a specific request to Facebook's Photo Team telling it to remove the photo comparison information.

Facebook touted the advantages of the new software on its blog: "Now if you upload pictures from your cousin's wedding, we'll group together pictures of the bride and suggest her name. Instead of typing her name 64 times, all you'll need to do is click 'Save' to tag all of your cousin's pictures at once. By making tagging easier than before, you're more likely to know right away when friends post photos. We notify you when you're tagged, and you can untag yourself at any time. As always, only friends can tag each other in photos."[49]

But what Facebook doesn't mention are the inherent dangers that come with this new ease of tagging photos. Take Facebook's wedding example. Perhaps you were tagged in a group reception photo early on in the night. But the facial recognition software also recognized you much later in the background of reception photo 62, chugging a Jägerbomb; 63, kissing a bridesmaid; and 64, vomiting on the dance floor. Now your drunken revelry has been tagged, you've been identified in the picture, and it's been posted prominently on your Facebook page. If your friend uploading the photos had to tag the photos without the help of facial recognition, your friend would have to first recognize your face in the background and then manually tag you in the photo while choosing to ignore the embarrassment, marital strife, or employment problems identifying you could cause.

And Facebook does not indicate in its privacy policy whether application developers, the government, or other third parties will be able to access or use this biometric data.[50] Instead of embracing informed consent and control, Facebook makes it extremely difficult to remove yourself from the unprecedented database containing 60 billion photos.[51]

In Germany in August 2011, the Hamburg Data Protection Authority[52] demanded that Facebook delete its database of biometric data on its citizens, which it believes violates European Union and German laws by storing data without the express consent of users.[53] Although Facebook insists that its facial recognition technology complies with EU data protection laws,[54] the Article 29 Working Party, which advises the European Union on its privacy policies, has expressed concern that Facebook's photo tag suggest feature is activated by default, with Facebook suggesting that a person be posted in a photo without obtaining his or her prior consent.[55]

When social networks make changes in policies without giving users adequate notice or control, individuals and policy advocates such as the Electronic Privacy Information Center seek justice at the Federal Trade Commission. In March 2010, the FTC filed a complaint against Google for unfair and deceptive trade practice in initiating Google Buzz. In response Google signed a consent agreement, promising to implement what *Legal Times* blogger Jenna Greene referred to as an "unprecedented privacy program."[56] The settlement requires Google to (1) not misrepresent the extent to which it protects users' privacy, (2) to give notice to users when their information will be disclosed to a third party and obtain "express affirmative consent" before sharing, and (3) to establish a comprehensive privacy program to address the privacy risks of Google's new and existing products and services and to report regularly to the FTC about its compliance in creating a new privacy program.[57] In a related civil lawsuit, Google settled consumer complaints about Google Buzz by agreeing to put $8.5 million into an independent fund to support privacy organizations.[58]

In the private lawsuit against Facebook challenging the Beacon program, Facebook agreed to establish a $9.5 million "settlement fund."[59] The settlement fund will be used to pay the attorney fees and trial expenses and $41,500 to the named plaintiffs—Sean Lane will get $15,000.[60] The remaining money in the settlement fund was to be used by Facebook to form a nonprofit organization, the Privacy Foundation, that will fund projects designed to increase awareness of "online privacy, safety, and security."[61] But one of the plaintiffs, Ginger McCall, has appealed the settlement, arguing that Facebook would retain an unacceptable level of control over the organization receiving the settlement funds.[62] "It puts the fox in charge of the henhouse," said Ginger.[63]

The facial recognition program met with similar resistance. On June 10, 2011, the Electronic Privacy Information Center, the Center for Digital Democracy, Consumer Watchdog, and the Privacy Rights Clearinghouse filed a complaint with the Federal Trade Commission, urging the agency to investigate Facebook's use of facial recognition software and biometric data, to require Facebook to cease collection and use of biometric data without affirmative opt-in consent, and to limit the disclosure of user information to third parties.[64] The complaint notes that facial recognition software has been used elsewhere to identify political dissidents.

The Chinese government is currently building a massive network of hidden cameras and facial recognition software to monitor its citizens in public spaces.[65] Currently, Chinese law requires that all internet cafés and nightclubs have secret cameras installed that send images directly to police stations, and facial recognition software is being used to process these images.[66]

In 2011, Carnegie Mellon researchers found that by combining facial recognition with data mining, they could predict the first five digits of a person's Social Security number from a webcam picture.[67] Those same researchers were able to use facial recognition software to match one in ten pseudonymous photos on dating sites to the real names from social network sites.

When the Department of Homeland Security proposed to initiate facial recognition technology in the United States, it met with such significant resistance that the program was scrapped. But the citizens of Facebook Nation were put into such a program by default. The EPIC complaint points out that the increasing amalgamation of data about people has led to increased identity theft and has also threatened the fundamental right of privacy, jeopardizing the opportunities to secure employment, insurance, and credit.

Under our Social Network Constitution, you should be told in advance what was going to happen to you on a social network and told in advance about any changes. No change should be possible to the privacy terms unless you affirmatively opt in. And, to enhance your right of control, no third party should be able to undertake data collection about you (whether through cookies, web beacons, deep packet inspection, or other techniques) on another entity's website. Rather than having to fight off third parties' cookies one by one by opting out, the entire process should be opt-in. Third parties (such as data aggregators and behavioral marketing firms) should have to place their invitations for you to opt in someplace else on the Web, rather than on social networks or your favorite websites.

The current practice of data aggregators would need to change immediately under our Social Network Constitution. The installation of cookies, Flash cookies, web beacons, and deep packet inspection would be impermissible without our specific consent based on adequate disclosure. Such a result would be required due to our substantive and procedural rights.

Currently, people's rights on Facebook shift at the whim of Mark Zuckerberg. The Federal Trade Commission is opening one investigation after another to try to keep up with each new complaint about the expansion in what Facebook makes public. But the entire enterprise lacks a governing set of principles. By adopting a Social Network Constitution that includes protections for due process, we can provide a template for individuals, institutions, websites, social networks, data aggregators, and government agencies to do what is right for the people of Facebook Nation.

14

Slouching Towards a Constitution

Cory Doctorow, co-editor of Boing Boing, who describes himself as a techno-utopian, surprised people when he took the floor at a TEDx conference and suggested the need for greater ways of protecting online privacy. He issued a call to action for people to use strategies and technologies to ensure the techno-privacy of themselves and their children.

Doctorow suggested that we've been going about our attempts to protect fundamental values in a wrong-headed way. When parents use software and other measures to monitor their children's online presence, "it trains kids to believe that surveillance of every move on the internet is a legitimate and proportionate thing for authority figures to do. It teaches them to systematically undervalue their privacy even before they reach Facebook. It in fact grooms them to accept total involuntary disclosure of their social intellectual life as normal and good."[1]

Doctorow proposes a radical change: "Let's turn our libraries, schools and other institutions into islands of network privacy best practices. Let's teach our kids to encrypt everything they do on the internet. . . . Teach kids to choose the best products for their privacy, even when big companies would prefer them not installing that software on their phone or tablet or laptop. This won't automatically make Facebook or its competitors stop selling our privacy. But it will legitimize the use of privacy tools and lay the ground for a future where 'why do you need to know this?' is the default position when someone asks our kids to disclose information over the network. I'm a father and I'm a teacher and I want to inhabit a world where networks continue to enhance our ability to work together without commodifying all of our relationships and committing us to these irrevocable punitive one-sided transparency disclosures."[2]

Consumer demand—or a Social Network Constitution or other policy requiring attention to fundamental rights—could foster the development of new technologies that respect privacy and individual choice. One step in this direction was taken by Google when it launched Google+ (Google Plus), a social network service that allows users to sort contacts into groups with separate sharing settings. For example, users may separate acquaintances from friends and employers from girls-night-out party groups. For each item that is posted (including photos and posts), the Google+ user selects the groups allowed to view it.[3] Users may also view their Google+ page as a particular group sees it, to ensure that content is being limited as desired.[4] "In real life, we have walls and windows and I can speak to you knowing who's in the room, but in the online world, you get to a 'Share' box and you share with the whole world," explained Bradley Horowitz, a vice president for product management at Google.[5] Google+ is also defaulted to conceal geographic location information for uploaded photos.[6] Google's marketing strategy was to assert that its product offered greater privacy and control than Facebook. But the extent to which greater protection is being provided will have to be monitored, as Google itself has been criticized in the past for its own privacy faux pas such as Google Buzz.

Los Angeles Times business columnist David Lazarus notes how the profit of companies, such as Spokeo, would dip if they were required to pay attention to people's fundamental rights such as by asking in advance if we want our information shared. "In the digital age, let's face it, there's no money in doing the right thing."[7] But as people begin to realize the risks of their posts being used against them, the demand for rights-protecting technologies and policies is growing. In 2009, 68.4% of people surveyed by TRUSTe indicated they "would use a browser feature that blocks ads, content and tracking code that doesn't originate from the site they're visiting."[8] And the majority of Americans favor politics that safeguard fundamental rights in the social network context: 68% of Americans oppose being "followed" on the Web, and 70% support the imposition of hefty fines for companies that collect or use someone's information without his or her consent.[9] Most people—92%—believe that websites and advertising companies should be required to delete all information stored about an individual if requested to do so.[10] Rather than people growing used to an abridgment of their fundamental rights—as Mark Zuckerberg posited—a 2010 Pew study reports that internet users in their late teens to early thirties are actually more likely to be concerned with their online privacy than are older users.[11] In late 2011, after the launch of Google+, even Facebook felt compelled to offer greater privacy safeguards, such as the creation of separate lists so that a person could share posts or photos with a subset of friends—perhaps showing drinking photos to best buddies and not an employer.

Around the world, policymakers have started to protect fundamental rights in the context of social networks and data aggregators. "Beginning with Germany in the mid-1980s, the Europeans moved towards a much broader concept of information privacy rights, leaving behind the Warren/Brandeis right to privacy," Viktor Mayer-Schönberger notes in *Delete: The Virtue of Forgetting in the Digital Age*. "If the original information privacy rights were more focused on the question of individual consent, information privacy is now seen as an individual's right to shape her participation in society."[12]

The European Union has established privacy laws covering data collection, storage, and dissemination.[13] In Europe, when a person's personal data is collected, the parties responsible for the collection are required to inform the person who they are, why they collected it, and for whom it was collected.[14] If entities use a person's data in any way, they are required to supply a copy of the data to the person in an "intelligible form" along with all the available information they have about the source of that data. If any part of the data is inaccurate or unlawfully processed, the person has the right to ask that they make a correction, a deletion, or completely erase the data. Should there be an inaccuracy, the person has the right to make the data collector notify all third parties who have seen the inaccurate data.

Most previous attempts at creating a bill of rights for the internet have focused on freedom from government oversight. In 1996, when John Perry Barlow wrote "A Declaration of the Independence of Cyberspace," his enthusiasm about the benefits of the internet was evident. His missive began: "Governments of the Industrial World, you weary giants of flesh and steel, I come from Cyberspace, the new home of Mind. On behalf of the future, I ask you of the past to leave us alone. You are not welcome among us. You have no sovereignty where we gather."

Even today, Jeff Jarvis (director of the interactive journalism program at City University of New York and a popular blogger) focuses his proposed Bill of Rights in Cyberspace on access and freedom.[15] But other commentators interested in governance of the internet have recognized the need for protections for users as well as the freedom to express themselves. In 2010, the attendees of the 21st Annual Computers, Freedom, and Privacy Conference came to the conclusion that "It's Time for a Social Network Users' Bill of Rights."[16] The proposed Bill of Rights would support not only values such as freedom of speech but also self-protection through privacy-enhancing techniques. However, the proposal from the Computers, Freedom, and Privacy Conference would apply only to the relationship between users and social network sites by modifying terms of service.[17] Our Social Network Constitution would apply more broadly to governments, social networks, other entities that use the internet, and to users themselves.

Our Social Network Constitution would ensure that all individuals have the opportunity to connect and freely associate without discrimination. It will give individuals control over their information, place, feelings, and image. It will ensure that the judicial system is able to fairly and appropriately administer justice. And, built into our Constitution, will be a mechanism protecting these rights.

Citizens of Facebook Nation are not the only ones who will benefit from the protections of a Social Network Constitution. In fact, the social networks themselves will benefit in the long run from the protection of individual rights and the establishment of a mechanism that will enforce these rights. Freedom of expression is a value that is coveted in our society. But if the price of exerting this right is losing a child or a chance at a fair trial, then people will stop expressing themselves on social networks and those networks will fade away.

Some people may oppose a Social Network Constitution on the grounds that social networks must be free to do whatever they want with no possibility of regulation. Imagine if we'd applied that argument to other technologies when they first came into use, such as cars. You wouldn't be able to seek justice if someone ran over your loved one in a car. You'd have no recourse if a manufacturer designed a car with faults such as unexpected acceleration or a gas tank that exploded on low level impact. What if there were no rules of the road, even determining which side of the street people should drive on? Imagine how many more lethal accidents there would be.

Facebook and other social networks provide fertile soil for personal growth — developing second selves and exploring individual identities. But without some regulation of the Web's Wild West, the value of participating in a social network will be undermined. "Will our children be outspoken in online equivalents of school newspapers if they fear their blunt words might hurt their future career?" asks Viktor Mayer-Schönberger in *Delete: The Virtue of Forgetting in the Digital Age*. "Will we protest against corporate greed or environmental destruction if we worry that these corporations may in some distant future refuse doing business with us? Just the thought that they might not, may constrain our willingness to act as consumers, let alone as citizens."[18]

Frustration with social networks' abridgment of fundamental rights is beginning to filter through the user community, technology design forums, and even the worldwide regulatory community. Applying fundamental values to social networks to create a Constitution for the Web could protect, online, key rights that people enjoy offline. Now is the time for the citizens of Facebook Nation, policymakers, and the social networks themselves to rally around a Social Network Constitution.

The Social Network Constitution

We the people of Facebook Nation, in order to form a more Perfect Internet, to protect our fundamental rights and freedoms, to explore our identities, dreams, and relationships, to safeguard the sanctity of our digital selves, to ensure equal access to technology, to lessen discrimination and disparities, and to promote democratic principles and the general welfare, declare these truths to be self-evident:

1. THE RIGHT TO CONNECT.

The right to connect is essential for individual growth, political discourse, and social interchange. No government shall abridge the right to connect, nor shall a government monitor exchanges over the internet or code them as to sources or content.

2. THE RIGHT TO FREE SPEECH AND FREEDOM OF EXPRESSION.

The right to free speech and freedom of expression shall not be abridged (and an individual shall have the freedom to use a pseudonym), as long as the speech does not incite serious, imminent harm nor defame a private individual. Employers and schools shall be prohibited from accessing social network pages or taking adverse actions against people based on what they express or disclose on a social network, except in cases of imminent harm to another individual.

3. THE RIGHT TO PRIVACY OF PLACE AND INFORMATION.

The right to privacy in one's social networking profiles, accounts, related activities, and data derived therefrom, shall not be abridged. The right to privacy includes the right to security of information and security of place. Regardless of active security settings or an individual's efforts to guard his or her digital self, social networks are private places.

4. THE RIGHT TO PRIVACY OF THOUGHTS, EMOTIONS AND SENTIMENTS.

Social networks provide a place for individuals to express themselves and to grow. A person's thoughts, emotions, and sentiments—and his or her characterization by others—shall not be used against him or her by social institutions, governments, schools, employers, insurers, or courts.

5. THE RIGHT TO CONTROL ONE'S IMAGE.

Each individual shall have control over his or her image from a social network, including over the image created by data aggregation. A person's image may not be used outside a social network for commercial or other purposes without his or her consent, nor shall it be used online for commercial or other gain without his or her consent.

6. THE RIGHT TO FAIR TRIAL.

Evidence from social networks may only be collected for introduction in a criminal trial if there is probable cause and a warrant has been issued. Evidence from social network sites may not be collected for or introduced in civil cases unless the activity at issue occurred on social networks (such as defamation, extortion, invasion of privacy, or jury tampering). Evidence from social networks may only be introduced at trial if it is directly relevant to the crime or civil action charged and the probative value outweighs the prejudicial value, the evidence is relevant, the evidence is properly authenticated, and the evidence otherwise complies with all rules of civil and criminal procedure. In custody cases, social network information should be admitted only if it provides direct evidence of past or potential harm to the child.

7. THE RIGHT TO AN UNTAINTED JURY.

Jurors shall decide cases based on the evidence presented in court and not information or inferences acquired from social networks, search queries, or other sources.

8. THE RIGHT TO DUE PROCESS OF LAW AND THE RIGHT TO NOTICE.

An individual is entitled to due process, which consists of advance notice and the ability to control, correct, and delete the individual's online information. No information shall be collected or analyzed without advance notification and consent of the individual. That notification shall include an explanation of the specific use and purpose for the collection and analysis of that information. There shall be a warning about possible repercussions of giving consent for the collection of that particular information. Access to a social network shall not be denied based on a decision not to consent to the collection, analysis, or dissemination of information. An individual shall have the right to know what entities are in possession of or are using that individual's information and he or she shall have a right to gain access to and obtain a copy of all such information regarding him or her.

9. FREEDOM FROM DISCRIMINATION.

No person shall be discriminated against based on his or her social network activities or profile, nor shall an individual be discriminated against based on group data aggregation rather than on characteristics of that particular individual, unless the social network activities provide direct evidence of a crime or tort.

10. FREEDOM OF ASSOCIATION.

People shall have freedom of association on social networks and the right to keep their associations private.

ACKNOWLEDGMENTS

For years, because of what I knew about computer security (or rather the lack thereof), I refused to buy anything on the Web. Not an airline ticket. Not even a book from Amazon.com. Then one of my research assistants wanted a specific birthday present that could only be purchased on stupid.com, a website that sells things like a flying alarm clock, a book of seriously bad baby names, fetus-shaped soap on a rope, and Freudian slippers. I made the purchase and realized that, even though I was a decently paid, hopefully well-respected law professor, my entire digital profile was now that of a stupid.com user.

Today, it's impossible to stay off the grid. In La Jolla, I had coffee with Ted Waitt, founder of Gateway Computers. He confessed that he was opposed to the data collection that occurs when a person uses a grocery chain savings card. But even as a billionaire, he couldn't resist taking advantage of the savings by using the card.

Now my preferences are revealed in thousands of ways by the Lori Andrews of the Web. From my Facebook profile to the website for my books, from my credit card purchases to my decision to check job postings, facts about me are floating through the ether and could be used against me. A stalker found me with just a quick glance at my book-signing schedule. Now he can learn my unlisted phone number and my home address through a free data aggregator. And unlike the grocery savings card, which I could choose not to use, I am not even informed about—let alone allowed to control—what information is made available about me.

As I've written this book, I've been fascinated by how my profile on the Web has changed. When I was doing research about the virtual Texas deputies who sat in pubs and digitally patrolled the border, the following ads popped up on my computer: "Designer Heels Only $39." "Newegg Business Store: Computers, Office Equipment, Office Supplies, Software and more!" And, "Earn a Homeland Security Degree & Become a Border Patrol Agent Today!"

Yeah, that's me. A Christian Louboutin–wearing, software-wielding border cop.

And I shudder to think of all the misconceptions I've fostered in the digital profiles of the lawyers, law students, and other research assistants who have come along with me on this journey across the Facebook Nation. Hats off to Jen Acker,

Sarah Blenner, Molly Brown, Robert Ennesser, Amanda Fraerman, Daniel Hantman, Kayla Kostelecky, Jake Meyer, Sarah Nelson, Elizabeth Raki, Cynthia Sun, and Keith Syverson. They've bravely marched into web searches using the suspect Homeland Security terms; found troubling online photos; investigated cyberharassment; sought out the data that aggregators have collected about them; and found, read, and analyzed hundreds of cases, statutes, and studies about social networks and their technological predecessors. I couldn't have done this book without their help, the advice of Emily Barney, and the amazing aid of the students in my Law of Social Networks seminar—Brandon Brooks, Samuel Coe, Alexis Crawford, Ashley Crettol, Alyssa Graber, Michelle Green, Jaclyn Hilderbrand, Richard Komaiko, Jeremiah Lewellen, Rachel Mercer, Elizabeth Meyer, Lauren Ortega, Oscar Rivera, Gabriela Sapia, William Saranow, and Mark Silverman.

I also owe a deep debt of gratitude to William Stubing and the Greenwall Foundation for funding my analysis of health information on social networks and to my colleagues at Chicago-Kent College of Law, some of whom pioneered the field of internet law. A special thanks to law professors Richard Warner, Hank Perritt, and Ron Staudt, pillars in cyberspace, who couldn't have been more gracious and helpful as I entered the field. Once again, Dean Hal Krent made himself available to read chapters and give sage advice about the direction of my project. A 40-professor, cross-disciplinary social networks group, headed by psychologist Ellen Mitchell at the Illinois Institute of Technology, helped provide the context for my work. The people who lived through the cases highlighted in the book, including Cynthia Moreno, Kenneth Zeran, and attorneys Eric Goldman, Jennifer Lynch, Laura Pirri, Dennis Riordan, and Julie Samuels, generously shared their stories with me.

And I couldn't face any task in my life—including writing a book—without the aid and comfort and editing advice of those dearest to me. Thanks once again to Christopher Ripley, Lesa Andrews, Felice Batlan, Francis Pizzulli, Clem Ripley, Jim Stark, and Darren Stephens. And to Amanda Urban and Emily Loose for believing in this project and always asking the most provocative questions.

NOTES

1. Facebook Nation

1. Nicholas Carlson, "Mark Zuckerberg Goes to England, Meets the Prime Minister," June 21, 2010, www.sfgate.com/cgi-bin/article.cgi?f=/g/a/2010/06/21/businessinsider-mark-zuckerberg-goes-to-england-meets-the-prime-minister-2010-6.DTL; Ian Burrell, "Power Profile," Feb. 15, 2011, www.gq-magazine.co.uk/comment/articles/2011-02/15/gq-comment-business-profile-joanna-shields-facebook; Number10Gov, "PM and Facebook Co-Founder Mark Zuckerberg," July 8, 2010, www.youtube.com/watch?v=b5Bbzi7s1Ko.

2. Number10Gov, "PM and Facebook Co-Founder Mark Zuckerberg."

3. "What is the G8?," G20-G8 France 2011, www.g20-g8.com/g8-g20/g8/english/what-is-the-g8-/what-is-the-g8-/what-is-the-g8.847.html.

4. Mark Duell, "That's Some Social Network! Facebook CEO Mark Zuckerberg Schmoozes with World Leaders at G8 Internet Session," May 27, 2011, www.dailymail.co.uk/news/article-1391380/G8-summit-2011-Facebook-CEO-Mark-Zuckerberg-schmoozes-world-leaders.html.

5. Michael Scherer, "Obama and Twitter: White House Social-Networking," May 6, 2009, www.time.com/time/politics/article/0,8599,1896482,00.html.

6. Ian Bogost, "Ian Became a Fan of Marshall McLuhan on Facebook and Suggested You Become a Fan Too," in *Facebook and Philosophy: What's on Your Mind?*, ed. D. E. Wittkower (Chicago: Open Court Publishing Company, 2010), 21–32.

7. Noah Shachtman, "Marines Ban Twitter, MySpace, Facebook," Aug. 3, 2009, www.wired.com/dangerroom/2009/08/marines-ban-twitter-myspace-facebook/.

8. Barbara Ortutay, "Sony Says Stolen PlayStation Credit Data Encrypted," April 28, 2011, http://abcnews.go.com/Technology/wireStory?id=13477769.

9. Amber Corrin, "DOD Rethinking Social-Media Access," Federal Computer Week, Aug. 3, 2009, http://fcw.com/Articles/2009/08/03/DOD-rethinking-social-media-access.aspx; Noah Schactman, "Military May Ban Twitter, Facebook as Security 'Headaches,'" July 30, 2009, www.wired.com/dangerroom/2009/07/military-may-ban-twitter-facebook-as-security-headaches/.

10. "The iPhone Goes to War," The New York Times Bits blog, Dec. 16, 2009, http://bits.blogs.nytimes.com/2009/12/16/the-iphone-goes-to-war/.

11. Tom Kaneshige, "U.S. Military Will Battle Test the iPhone," Dec. 14, 2010, http://advice.cio.com/tom_kaneshige/14722/u_s_military_will_battle_test_the_iphone.

12. Tim Devaney, "Soldiers on Battlefield Turn Apps into Arms," Jan. 24, 2011, www.washingtontimes.com/news/2011/jan/24/soldiers-on-battlefield-turn-apps-into-arms/?page=all#pagebreak.

13   Deputy Secretary of Defense, "Directive-Type Memorandum DTM 09-026—Responsible and Effective Use of Internet-based Capabilities," Feb. 25, 2010.

14   Bob Brewin, "Army Confirms Battlefield Smartphones Tests Began in December," March 29, 2011, www.nextgov.com/nextgov/ng_20110329_6868.php; Joe Gould and Michael Hoffman, "Army Sees Smart Phones Playing Important Role," Dec. 12, 2010, www.armytimes.com/news/2010/12/army-smart-phones-for-soldiers-121210w/.

15   Tim Devaney, "Soldiers on Battlefield Turn Apps into Arms," Jan. 24, 2011, www.washingtontimes.com/news/2011/jan/24/soldiers-on-battlefield-turn-apps-into-arms/?page=all#pagebreak.

16   Department of Homeland Security, "Terrorist Use of Social Networking Sites Facebook Case Study," Dec. 5, 2010, http://publicintelligence.net/ufouoles-dhs-terrorist-use-of-social-networking-facebook-case-study/.

17   Ibid.

18   *State v. McGuire*, 16 A.3d 411, 423–424 (N.J. Super. Ct. App. Div. 2011).

19   Jennifer Dobner, "Got a Cute 'Hostage' Huh: Wanted Man Updates Facebook Status During 16-Hour Stand-off," June 22, 2011, www.smh.com.au/technology/technology-news/got-a-cute-hostage-huh-wanted-man-updates-facebook-status-during-16hour-standoff-20110622-1ge63.html.

20   Jen Doll, "Jason Valdez Facebooks Holding a Woman Hostage in 16-Hour Police Standoff," June 22, 2011, http://blogs.villagevoice.com/runninscared/2011/06/jason_valdez_fa.php.

21   David Kirkpatrick, *The Facebook Effect* (New York: Simon & Schuster, 2010), 205.

22   *NAACP v. Alabama*, 357 U.S. 449, 460 (1958).

23   Ibid. at 462.

24   Jim Meyer, "The Officer Who Posted Too Much on MySpace," *The New York Times*, March 10, 2009, at A24, www.nytimes.com/2009/03/11/nyregion/11about.html.

25   Jackie Cohen, "ALERT: Job Screening Agency Archiving All Facebook," June 20, 2011, www.allfacebook.com/alert-job-screening-agency-archiving-all-facebook-2011-06.

26   Paolo Cirio and Alessandro Ludovico, "Face to Facebook," 2011, www.transmediale.de/content/face-facebook.

27   "How We Did It," www.face-to-facebook.net/how.php.

28   Ryan Singel, "'Dating' Site Imports 250,000 Facebook Profiles, Without Permission," Feb. 3, 2011, www.wired.com/epicenter/2011/02/facebook-dating/.

29   Paolo Cirio and Alessandro Ludovico, "Face to Facebook," 2011, www.transmediale.de/content/face-facebook.

30   U.S. Department of Homeland Security, "Privacy Impact Assessment for the Office of Operations Coordination and Planning: Publicly Available Social Media Monitoring and Situational Awareness Initiative Update," Jan. 6, 2011, www.dhs.gov/xlibrary/assets/privacy/privacy_pia_ops_publiclyavailablesocialmedia_update.pdf.

31   Jennifer Lynch, "Applying for Citizenship? U.S. Citizenship and Immigration Wants to be Your 'Friend,'" Oct. 12, 2010, www.eff.org/deeplinks/2010/10/applying-citizenship-u-s-citizenship-and.

32   "Social Networking Sites and their Importance to FDNS," U.S. Citizenship and Immigration Services, May 2008, https://www.eff.org/files/filenode/social_network/DHS_CustomsImmigration_SocialNetworking.pdf.

33   "Investigate your MP's Expenses," http://mps-expenses.guardian.co.uk/.

34   "About BlueServo," www.blueservo.net/about.php; Brandi Grissom, "Border Cameras Produce Little in Two Years," April 20, 2010, www.texastribune.org/texas-mexico-border-news/border-cameras/border-cameras-produce-little-in-two-years/.

35   "About BlueServo," www.blueservo.net/about.php; John Burnett, "A New Way to Patrol the Texas Border: Virtually," Feb. 23, 2009, www.npr.org/templates/story/story.php?storyId=101050132.

36   "Virtual Stake Outs—Live Border Cameras," www.blueservo.net/vcw.php.

37   Burnett, "A New Way to Patrol the Texas Border: Virtually."

38   John Sutter, "Guarding the U.S.-Mexico Border, Live from Suburban New York," March 12, 2009, http://articles.cnn.com/2009-03-12/tech/border.security.cameras.immigration_1_us-mexico-border -southern-border-illegal-immigration?_s=PM:TECH.

39   "Virtual Border System Ineffective, Out of Cash," July 16, 2009, www.homelandsecuritynewswire .com/virtual-border-system-ineffective-out-cash?page=0,0.

40   Claire Prentice, "Armchair Deputies Enlisted to Patrol US-Mexico Border," Dec. 26, 2009, http:// news.bbc.co.uk/2/hi/8412603.stm.

41   Grissom, "Border Cameras Produce Little in Two Years."

42   "Virtual Border System Ineffective, Out of Cash."

43   Prentice, "Armchair Deputies Enlisted to Patrol US-Mexico Border."

44   "Virtual Border System Ineffective, Out of Cash."

45   Prentice, "Armchair Deputies Enlisted to Patrol US-Mexico Border."

46   "About BlueServo," www.blueservo.net/about.php.

47   Matthew Moore, "YouTube 'Cat Torturer' Traced by Web Detectives," Feb. 17, 2009, www.telegraph .co.uk/news/worldnews/northamerica/usa/4678878/YouTube-cat-torturer-traced-by-web-detectives .html.

48   "Kenny Glenn," http://ohinternet.com/Kenny_Glenn.

49   Carl Campanile, "Dem Pol's Son was 'Hacker,'" Sep. 19, 2008, www.nypost.com/p/news/politics /item_IJuyiNfQAkPKvZxRXvSEyJ;jsessionid=E00BF6768C23A24BD95F4E906DDD74B7; Bill Poovey, "David Kernell, Palin E mail Hacker, Sentenced to Year in Custody," Nov. 12, 2010, www.huffingtonpost .com/2010/11/12/david-kernell-palin-email_n_782820.html.

50   Megan Sayers, "Social Media and the Law: Police Explore New Ways to Fight Crime," March 30, 2011, http://thenextweb.com/socialmedia/2011/03/30/social-media-and-the-law-police-explore-new -ways-to-fight-crime/.

51   Laura Saunders, "Is 'Friending' in Your Future? Better Pay Your Taxes First," *The Wall Street Journal*, Aug. 27, 2009, at A2, http://online.wsj.com/article/SB125132627009861985.html.

52   Miguel Helft, "Google Uses Web Searches to Track Flu's Spread," *The New York Times*, Nov. 11, 2008, at A1, www.nytimes.com/2008/11/12/technology/internet/12flu.html.

53   Mark Sullivan, "How Will Facebook Make Money?," June 2010, www.pcworld.com/article/198815 /how_will_facebook_make_money.html.

54   Ibid.

55   Justin Smith, "10 Powerful Ways to Target Facebook Ads Every Performance Advertiser Should Know," July 27, 2009, www.insidefacebook.com/2009/07/27/10-powerful-ways-to-target-facebook-ads -that-every-performance-advertiser-should-know/.

56   Webmaster BNXS, "Huge Disparity in Facebook's Ads Display and Ads Revenue," Jan. 19, 2011, http://bnxs.com/huge-disparity-in-facebook%E2%80%99s-ads-display-and-ads-revenue/.

57   Sullivan, "How Will Facebook Make Money?."

58   Deborah Liu, "The Next Step for Facebook Credits," Jan. 24, 2011, http://developers.facebook.com /blog/post/451.

59   Kirkpatrick, *The Facebook Effect*, 232.

60   "About Spokeo," www.spokeo.com/blog/about/; Complaint, Request for Investigation, Injunction, and Other Relief to the Federal Trade Commission from the Center for Democracy and Technology at ¶ 2, *In the Matter of Spokeo* (June 30, 2010).

61   "About Spokeo," www.spokeo.com/blog/about/.

62   *See generally*, www.spokeo.com; David Lazarus, "You Won't Find Spokeo Founder Included on His 'People Search' Site," *Los Angeles Times*, June 8, 2010, at 1, http://articles.latimes.com/2010/jun/08 /business/la-fiw-lazarus-20100608.

63   "Spokeo home page," www.spokeo.com.

64   "Plans and Pricing," www.spokeo.com/plans.

65   Ibid.

66   "Frequently Asked Questions: What Is a Spokeo Premium Subscription?," www.spokeo.com/blog/help/.

67   "Plans and Pricing," www.spokeo.com/plans.

68   First Amended Complaint at ¶ 16, *Robins v. Spokeo*, No. 10-CV-05306 (C.D. Cal. Feb. 16, 2011).

69   15 U.S.C. §1681a, (West, Current through PL 112-24).

70   "Spokeo home page," www.spokeo.com.

71   "About Spokeo," www.spokeo.com/blog/about/; Complaint, Request for Investigation, Injunction, and Other Relief to the Federal Trade Commission from the Center for Democracy and Technology at ¶ 13, *In the Matter of Spokeo* (June 30, 2010).

72   Complaint at ¶¶ 21-23, *Robins v. Spokeo*, No. 10-CV-05306 (C.D. Cal. July 20, 2010).

73   Order Granting in Part and Denying in Part Defendant's Motion to Dismiss Plaintiff's First Amended Complaint, *Robins v. Spokeo*, No. 10-CV-05306 (C.D. Cal. May 11, 2011).

74   Ibid.; *But see*, Order Correcting Prior Ruling and Finding Moot Motion for Certification, *Robins v. Spokeo*, 10-cv-05306 (C.D. Cal. Sep. 19, 2011).

75   "CR Investigates: Your Privacy for Sale," *Consumer Reports*, October 2006, at 41, www.nofixnopay .info/Your_privacy_for_sale.htm.

76   Deborah Pierce and Linda Ackerman, "Data Aggregators: A study of Data Quality and Responsiveness," PrivacyActivism.org, May 19, 2005, www.csun.edu/~dwm3265/IS312/DataAggregatorsStudy.pdf.

77   Ibid.; "CR Investigates: Your Privacy for Sale."

78   "CR Investigates: Your Privacy for Sale."

79   *People v. Klapper*, 902 N.Y.S.2d 305 (N.Y. Crim. Ct. 2010).

80   David Lazarus, "Forget Privacy; He Sells Your Data," *Los Angeles Times*, June 8, 2010, at A1, published online as "You Won't Find Spokeo Founder Included on His 'People Search' Site," http:// articles.latimes.com/2010/jun/08/business/la-fiw-lazarus-20100608.

81   Josh, "Facebook Disabled My Account," Jan. 3, 2008, http://scobleizer.com/2008/01/03/ive-been -kicked-off-of-facebook/.

82   Face to Facebook Press Release, April 7, 2011, www.ecopolis.org/category/art/.

83   Laura Allsop, "Art 'Hacktivists' Take on Facebook," Feb. 11, 2011, http://edition.cnn.com/2011 /WORLD/europe/02/11/artists.facebook.project/index.html?section=cnn_latest.

84   Face to Facebook Press Release, April 7, 2011, www.ecopolis.org/category/art/.

85   Matthew Fraser and Soumitra Dutta, "Barack Obama and the Facebook Election," Nov. 19, 2008, www .usnews.com/opinion/articles/2008/11/19/barack-obama-and-the-facebook-election; "Exit Polls: Obama Wins Big Among Young Minority Voters," Nov. 4, 2008, www.usnews.com/opinion/articles/2008/11/19 /barack-obama-and-the-facebook-election.

86   Prime Minister David Cameron, "PM Statement on Disorder in England," Aug. 11, 2011, www .number10.gov.uk/news/pm-statement-on-disorder-in-england/.

87   Tom Pettifor, Andrew Gregory and Josh Layton, "UK Riots: Untraceable BlackBerry Messenger Should Be Suspended, Claims Tottenham MP David Lammy," Aug. 20, 2011, www.mirror.co.uk /news/technology/2011/08/10/uk-riots-tottenham-mp-david-lammy-calls-on-blackberry-to-suspend -network-to-stop-rioters-organising-trouble-115875-23333287/.

88   Josh Halliday, "David Cameron Considers Banning Suspected Rioters from Social Media," Aug. 11, 2011, www.guardian.co.uk/media/2011/aug/11/david-cameron-rioters-social-media.

89   Ibid.

90   "The Crunchies Awards," UStream (Michael Arrington interviews Mark Zuckerberg), video from Terrence O'Brien, "Facebook's Mark Zuckerberg Claims Privacy is Dead," Jan. 11, 2010, www .switched.com/2010/01/11/facebooks-mark-zuckerberg-claims-privacy-is-dead/.

91   Kirkpatrick, *The Facebook Effect*, 203.

92   Stephen Gardbaum, "The 'Horizontal Effect' of Constitutional Rights," 102 *Michigan Law Review* 387 (2003).

2. George Orwell . . . Meet Mark Zuckerberg

1   Federal Trade Commission, "FTC Staff Report: Self-Regulatory Principles for Online Behavioral Advertising—Behavioral Advertising: Tracking, Targeting, & Technology," 2009 WL 361109 at 4 (February 2009).

2   Audience Science Press Release, "State of Audience Targeting Industry Study: 50% of Advertisers Set to Boost Spending on Audience Targeting in 2011," Jan. 11, 2011, www.audiencescience.com/uk /press-room/press-releases/2011/state-audience-targeting-industry-study-50-advertisers-set-boost-spen.

3   Internet Advertising Bureau, Internet Advertising Revenue Report, 2010 Full Year Results, April 2011, www.iab.net/insights_research/947883/adrevenuereport.

4   ComScore Press Release, "Americans Received 1 Trillion Display Ads in Q1 2010 as Online Advertising Market Rebounds from 2009 Recession," May 13, 2010, www.comscore.com/Press _Events/Press_Releases/2010/5/Americans_Received_1_Trillion_Display_Ads_in_Q1_2010_as _Online_Advertising_Market_Rebounds_from_2009_Recession.

5   Louise Story, "F.T.C. to Review Online Ads and Privacy," *The New York Times*, Nov. 1, 2007, at C1, www.nytimes.com/2007/11/01/technology/01Privacy.html?ref=technology.

6   Nicholas Carlson, "Facebook Expected to File for $100 Billion IPO This Year," June 13, 2011, www .businessinsider.com/facebook-ipo-could-come-in-q1-2012-after-october-filing-cnbc-reports-2011-6.

7   Stephanie Reese, "Quick Stat: Facebook to Bring in $4.05 Billion in Ad Revenues This Year," April 26, 2011, www.emarketer.com/blog/index.php/tag/facebook-ad-revenue/.

8   Cory Doctorow, Talk at TEDx Observer, 2011, http://tedxtalks.ted.com/video/TEDxObserver-Cory -Doctorow.

9   Ibid.

10   Complaint at 2, *Valentine v. NebuAd, Inc.*, No. C08-05113 TEH (N.D. Cal. Nov. 10, 2008); Karl Bode, "Infighting at ISPs over Using NebuAD," May 29, 2008, www.dslreports.com/shownews /Infighting-At-ISPs-Over-Using-NebuAD-94835; *Valentine v. NebuAd, Inc.*, 2011 WL 1296111 (N.D. Cal. 2011).

11   Complaint at 2, *Valentine v. NebuAd, Inc.*, No. C08-05113 TEH (N.D. Cal. Nov. 10, 2008).

12   John L. McKnight, Curriculum Vita, www.northwestern.edu/ipr/people/jlmvita.pdf.

13   Shirley Sagawa and Eli Segal, *Common Interest, Common Good: Creating Value Through Business and Social Sector Partnerships* (Boston: Harvard Business Press, 2000), 30.

14   D. Bradford Hunt, "Redlining," The Electronic Encyclopedia of Chicago, www.encyclopedia .chicagohistory.org/pages/1050.html.

15   Marcia Stepanek, "Weblining," April 3, 2000, www.businessweek.com/2000/00_14/b3675027.htm.

16   David Goldman, "These Data Miners Know Everything About You," Dec. 16, 2010, http://money
     .cnn.com/galleries/2010/technology/1012/gallery.data_miners/index.html.

17   Rowena Mason, "Acxiom: the Company That Knows if you Own a Cat or if You're Right-Handed,"
     April 27, 2009, www.telegraph.co.uk/finance/newsbysector/retailandconsumer/5231752/Acxiom-the
     -company-that-knows-if-you-own-a-cat-or-if-youre-right-handed.html.

18   Goldman, "These Data Miners Know Everything About You."

19   Ian Ayres, *Super Crunchers* (New York: Bantam Dell 2007), 134.

20   Complaint at 4, *U.S. v. ChoicePoint*, No. 06-CV-0198 (N.D. Ga. Jan. 30, 2006).

21   Ibid. at 4-6; Stipulated Final Judgment and Order for Civil Penalties, Permanent Injunction, and
     Other Equitable Relief, *U.S. v. ChoicePoint*, No. 06-CV-0198 (N.D. Ga. Feb. 10, 2006).

22   Marcia Savage, "LexisNexis Security Breach Worse Than Thought," April 12, 2005, www
     .scmagazineus.com/lexisnexis-security-breach-worse-than-thought/article/31977/; Toby Anderson,
     "LexisNexis Owner Reed Elsevier Buys ChoicePoint," Feb. 21, 2008, www.usatoday.com/money
     /industries/2008-02-21-reed-choicepoint_N.htm.

23   Chris Cuomo, Jay Shaylor, Mary McGuirt, and Chris Francescani, "'GMA' Gets Answers: Some
     Credit Card Companies Financially Profiling Customers," Jan. 28, 2009, http://abcnews.go.com
     /GMA/GetsAnswers/Story?id=6747461.

24   Eli Pariser, *The Filter Bubble: What the Internet Is Hiding from You* (New York: Penguin, 2011),
     164.

25   Aleecia M. McDonald and Lorrie F. Cranor, "Americans' Attitudes About Internet Behavioral
     Advertising Practices," *Proceedings of the 9th Workshop on Privacy in the Electronic Society* WPES,
     Oct. 4, 2010, 6.

26   "Consumer Reports Poll: Americans Extremely Concerned About Internet Privacy," Sep. 25, 2008,
     www.consumersunion.org/pub/core_telecom_and_utilities/006189.html.

27   Joseph Turow, Jennifer King, Chris Jay Hoofnagle, Amy Bleakley, and Michael Hennessy, "Contrary
     to What Marketers Say, Americans Reject Tailored Advertising and Three Activities That Enable It,"
     September 2009, at 3, www.ftc.gov/os/comments/privacyroundtable/544506-00113.pdf.

28   Julia Angwin and Tom McGinty, "Sites Feed Personal Details to New Tracking Industry," *The Wall
     Street Journal*, July 30, 2010, at A1, http://online.wsj.com/article/SB10001424052748703977004575
     393173432219064.html.

29   "Tracking The Companies That Track You Online," Dave Davies's interview with Julia Angwin, *Fresh
     Air*, Aug. 19, 2010, www.npr.org/templates/story/story.php?storyId=129298003; Julia Angwin, "The
     Web's New Gold Mine: Your Secrets," *The Wall Street Journal*, July 30, 2010, at W1, http://online
     .wsj.com/article/SB10001424052748703940904575395073512989404.html.

30   Mike Elgan, "Snooping: It's Not a Crime, It's a Feature," April 16, 2011, www.computerworld.com/s
     /article/print/9215853/Snooping_It_s_not_a_crime_it_s_a_feature.

31   Hal Berghel, "Caustic Cookies," 44 *Communications of the ACM*, at 19-20 (2001).

32   David M. Kristol, "HTTP Cookies: Standards, Privacy, and Politics," 1 *ACM Transactions on Internet
     Technology* 151, 154 (2001), http://arxiv.org/PS_cache/cs/pdf/0105/0105018v1.pdf.

33   Ibid.

34   Ibid.

35   *In re DoubleClick Inc. Privacy Litigation*, 154 F. Supp. 2d 497 (S.D.N.Y. 2001). Consolidated
     Amended Class Action Complaint, *In re DoubleClick Inc. Privacy Litigation*, No. 00-Civ-0641
     (S.D.N.Y. Mar. 26, 2000).

36   Dan Butler, "More Snooping Around on the Internet," www.thenakedpc.com/articles/v04/13/0413
     -04.html.

37   "Cookies and Privacy Demonstration," http://cyber.law.harvard.edu/ilaw/Privacy/Demo.html.

38   Network Advertising Initiative, "Web Beacons—Guidelines for Notice and Chaos," www
      .networkadvertising.org/networks/initiatives.asp.

39   Ibid.

40   Ibid.

41   Joshua Gomez, Travis Pinnick, and Ashkan Soltani, KnowPrivacy.org Report, U.C. Berkeley School
      of Information, June 1, 2009, at 8, http://knowprivacy.org/report/KnowPrivacy_Final_Report.pdf.

42   Richard M. Smith, "The Web Bug FAQ," Nov. 11, 1999, http://w2.eff.org/Privacy/Marketing/?.

43   Network Advertising Initiative, "Web Beacons—Guidelines for Notice and Chaos."

44   Gomez et al., KnowPrivacy.org Report, at 8.

45   Ibid.

46   Mike Eckler, "An Introduction to Cookies; How to Manage Them," March 16, 2011, www
      .practicalecommerce.com/articles/2653-An-Introduction-to-Flash-Cookies-How-to-Manage-Them.

47   John Herman, "What Are Flash Cookies and How Can You Stop Them?," Sep. 23, 2010, www
      .popularmechanics.com/technology/how-to/computer-security/what-are-flash-cookies-and-how-can
      -you-stop-them.

48   Ashkan Soltani, Shannon Canty, Quentin Mayo, Lauren Thomas, and Chris Jay Hoofnagle,
      "Flash Cookies and Privacy," Working Paper Series, August 2009, papers.ssrn.com/sol3/papers
      .cfm?abstract_id=1446862; "What are Local Shared Objects?," www.adobe.com/products
      /flashplayer/articles/lso/.

49   Soltani et al., "Flash Cookies and Privacy."

50   CaptainPC, "And That's the Way the Cookie Crumbles," Aug. 26, 2010, http://allnurses-central.com
      /general-blogs/s-way-cookies-500525.html.

51   Eckler, "An Introduction to Cookies; How to Manage Them."

52   Paul Ohm, "The Rise and Fall of Invasive ISP Surveillance," 2009 University of Illinois Law Review
      1417, 1439 (2009).

53   In re DoubleClick Inc. Privacy Litigation, 154 F. Supp. 2d 497 (S.D.N.Y. 2001).

54   "What Your Broadband Provider Knows About Your Web Use: Deep Packet Inspection and
      Communications Laws and Policies," Statement of Alissa Cooper, Chief Computer Scientist,
      Center for Democracy & Technology, before the House Committee on Energy and Commerce,
      Subcommittee on Telecommunications and the Internet, July 17, 2008, 7.

55   Ibid.

56   A nonprofit public interest organization that promotes "free expression and privacy in communication
      technologies." See www.cdt.org/about.

57   "What Your Broadband Provider Knows About Your Web Use," 4.

58   Ibid.

59   Jami Makan, "10 Things Facebook Won't Say," Jan. 10, 2011, www.smartmoney.com/spend
      /technology/10-things-facebook-wont-say-1294414171193/.

60   Emily Steel and Geoffrey A. Fowler, "Facebook in Privacy Breach," The Wall Street Journal, Oct.
      18, 2010, at A1, http://online.wsj.com/article/SB10001424052702304772804575558484075236968
      .html.

61   Ibid.

62   Ibid.

63   "Web Scraping Tutorial," March 7, 2009, www.codediesel.com/php/web-scraping-in-php-tutorial/.
      CodeDiesel is a web development journal.

64   Sean O'Reilly, "Nominative Fair Use and Internet Aggregators: Copyright and Trademark Challenges Posed by Bots, Web Crawlers and Screen-Scraping Technologies," 19 *Loyola Consumer Law Review* 273 (2007).

65   "Google Privacy FAQ," www.google.com/intl/en/privacy/faq.html#toc-terms-server-logs.

66   "Bing Privacy Supplement," January 2011, http://privacy.microsoft.com/en-us/bing.mspx.

67   "Google Privacy FAQ."

68   Omer Tene, "What Google Knows: Privacy and Internet Search Engines," 2008 *Utah Law Review* 1433, 1454 (2008).

69   "Yahoo! Privacy Policy," http://info.yahoo.com/privacy/us/yahoo/details.html.

70   Tene, "What Google Knows: Privacy and Internet Search Engines."

71   Michael Arrington, "AOL Proudly Releases Massive Amounts of Private Data," Aug. 6, 2006, http://techcrunch.com/2006/08/06/aol-proudly-releases-massive-amounts-of-user-search-data/.

72   Abdur Chowdhury, Email sent to SIG-IRList newsletter, Aug. 3, 2006, http://sifaka.cs.uiuc.edu/xshen/aol/20060803_SIG-IRListEmail.txt.

73   Declan McCullagh, "AOL's Disturbing Glimpse into Users' Lives," Aug. 7, 2006, http://news.cnet.com/AOLs-disturbing-glimpse-into-users-lives/2100-1030_3-6103098.html#ixzz1M56yaUU2.

74   Michael Barbaro and Tom Zeller, Jr., "A Face Is Exposed for AOL Searcher 4417749," *The New York Times*, Aug. 9, 2006, at A1, www.nytimes.com/2006/08/09/technology/09aol.html?pagewanted=all.

75   Ibid.

76   Tene, "What Google Knows: Privacy and Internet Search Engines."

3. Second Self

1   Deborah Copaken Kogan, "How Facebook Saved My Son's Life," July 13, 2011, www.slate.com/id/2297933/.

2   Josh Grossberg, "New Kid Helps Find Kid New Kidney," April 26, 2011, http://today.msnbc.msn.com/id/42770534/ns/today-entertainment/t/new-kid-helps-find-kid-new-kidney/; "Donnie Wahlberg Finds Sick Fan New Kidney Using Twitter," April 27, 2011, www.myfoxboston.com/dpp/news/local/donnie-wahlberg-finds-fan-kidney-using-twitter-20110427#ixzz1SfkAJfJx.; Stacy McCloud, "Music City Beat," July 16, 2011, http://fox17.com/newsroom/features/music-beat/videos/vid_706.shtml.

3   "About Us," www.patientslikeme.com/about.

4   "Patients," www.patientslikeme.com/patients.

5   "Sign up," www.patientslikeme.com/user/signup.

6   Julia Angwin and Steve Stecklow, "'Scrapers' Dig Deep for Data on Web," *The Wall Street Journal*, Oct. 12, 2010, at A1, http://online.wsj.com/article/SB10001424052748703358504575544381288117888.html. He was a resident of Sydney, Australia.

7   Ibid. Reportedly, unless it has permission, Nielsen now doesn't scrape websites that require an individual account for access.

8   Ibid.

9   "Patients," www.patientslikeme.com/patients/view/112762?patient_page=5.

10   Ibid.

11   45 C.F.R. §§ 160, 164; 42 U.S.C. §§ 300gg-51 et seq.; 29 U.S.C. 1182(b).

12   Emily Steel and Julia Angwin, "On the Web's Cutting Edge, Anonymity in Name Only," *The Wall Street Journal*, Aug. 4, 2010, at A1, http://online.wsj.com/article/SB10001424052748703294904575385532109190198.html.

13   Ibid.

14   Ian Ayres, *Super Crunchers* (New York: Bantam Dell, 2007), 134.

15   Eli Pariser, *The Filter Bubble: What the Internet Is Hiding from You* (New York: Penguin, 2011), 7.

16   "My Cluster," www.acxiom.com/Personicx_Cluster.aspx.

17   "Marketing Segmentation," PersonicX Interactive Wheel, www.acxiom.com/products_and_services /Consumer%20Insight%20Products/segmentation/Pages/index.html.

18   "My Cluster," www.acxiom.com/Personicx_Cluster.aspx.

19   Gary A. Hernandez, Katherine J. Eddy, and Joel Muchmore, "Insurance Weblining and Unfair Discrimination in Cyberspace," 54 *Southern Methodist University Law Review* 1953, 1965 (2001).

20   Ibid. at 1968.

21   Steel and Angwin, "On the Web's Cutting Edge, Anonymity in Name Only"; "Privacy," www .xplusone.com/privacy.php.

22   Ibid.

23   Marcia Stepanek, "Weblining," April 3, 2000, www.businessweek.com/2000/00_14/b3675027.htm.

24   Marcy Peek, "Passing Beyond Identity on the Internet: Espionage & Counterespionage in the Internet Age," 28 *Vermont Law Review* 91, 105 (2003).

25   Julia Angwin, "The Web's New Gold Mine: Your Secrets," *The Wall Street Journal*, July 30, 2010, at W1, http://online.wsj.com/article/SB10001424052748703940904575395073512989404.html.

26   Ibid.

27   Louise Story, "F.T.C. to Review Online Ads and Privacy," *The New York Times*, Nov. 1, 2007, at C1, www.nytimes.com/2007/11/01/technology/01Privacy.html?ref=technology.

28   Angwin, "The Web's New Gold Mine: Your Secrets."

29   Ibid.

30   Jim Edwards, "Why Google Took So Long to Pull Ads from Suicide Chat Group," Sep. 27, 2010, www.bnet.com/blog/advertising-business/why-google-took-so-long-to-pull-ads-from-suicide-chat -group/5956.

31   Ibid.

32   Jason Lewis, "Google Admits Cashing in on Suicide Pact Chatroom: Internet Giant Sold Ads on Web Pages Where Two British Strangers Arranged to Kill Themselves," Sep. 26, 2010, www.dailymail .co.uk/news/article-1315296/Google-admits-cashing-suicide-pact-chatroom-Internet-giant-sold -ads-web-pages-British-strangers-arranged-kill-themselves.html.

33   Ken Spencer Brown, "Google Web Search Is a Game-Changer in Advertising Field," *Investor's Business Daily*, July 16, 2007, at A1.

34   Lewis, "Google Admits Cashing in on Suicide Pact Chatroom."

35   Center for Digital Democracy and the U.S. Public Interest Research Group, "Supplemental Statement in Support of Complaint and Request for Inquiry and Injunctive Relief Concerning Unfair and Deceptive Online Marketing Practices," Nov. 1, 2007, at 29, *citing* "Google, Microsoft Top Nielsen/NetRatings Web Site Lists," *Website Zoom*, Sep. 19, 2007.

36   Faisal Laljee, "Subprime Mortgage Bust Could Create Ad Trouble for Google," Feb. 22, 2007, http:// seekingalpha.com/article/27736-subprime-mortgage-bust-could-create-ad-trouble-for-google.

37   Ibid.

38   Center for Digital Democracy and the U.S. Public Interest Research Group, "Supplemental Statement in Support of Complaint and Request for Inquiry and Injunctive Relief Concerning Unfair and Deceptive Online Marketing Practices," Nov. 1, 2007, at 29, *citing* Manny Fernandez, "Study Finds Disparities in Mortgages by Race," Oct. 15, 2007, www.nytimes.com/2007/10/15 /nyregion/15subprime.html?_r=1&ref=realestate&oref=slogin.

39   Ibid., *citing* "Sub Prime in Real Time," Oct. 11, 2007, http://pr-gb.com/index.php?option=com_content&task=view&id=29772&Itemid=9.

40   Ibid., *citing* Fernandez, "Study Finds Disparities in Mortgages by Race."

41   Ibid.

42   Jenny Anderson and Heather Timmons, "Why a U.S. Subprime Mortgage Crisis Is Felt Around the World," *The New York Times*, Aug. 31, 2007, at C1, www.nytimes.com/2007/08/31/business/worldbusiness/31derivatives.html, stating that the global financial crisis was "set off by problems with subprime mortgages; *see generally*, William Poole, "Causes and Consequences of the Financial Crisis of 2007–2009," 33 *Harvard Journal of Law and Public Policy* 421 (2010).

43   Katalina M. Bianco, "The Subprime Lending Crisis: Causes and Effects of the Mortgage Meltdown," May 2008, http://business.cch.com/bankingfinance/focus/news/Subprime_WP_rev.pdf.

44   Ibid.

45   *In re DoubleClick Inc. Privacy Litigation*, 154 F. Supp. 2d 497 (S.D.N.Y. 2001).

46   Laurie Petersen, "Microsoft $6B Deal Caps Watershed Month for Digital," May 21, 2007, www.mediapost.com/publications/index.cfm?fa=Articles.showArticle&art_aid=60652.

47   Stefan Berteau, "Facebook's Misrepresentation of Beacon's Threat to Privacy: Tracking Users Who Opt Out or Are Not Logged In," Nov. 29, 2007, http://community.ca.com/blogs/securityadvisor/archive/2007/11/29/facebook-s-misrepresentation-of-beacon-s-threat-to-privacy-tracking-users-who-opt-out-or-are-not-logged-in.aspx.

48   Complaint at ¶ 12, *In the Matter of Chitika, Inc.*, 2011 WL 914035 (F.T.C. March 14, 2011).

49   "Targeting," www.datranmedia.com/aperture/targeting/.

50   "Opt-Out Notice," http://rt.displaymarketplace.com/optout.html.

51   Stephen Shankland, "Adobe Tackling 'Flash Cookie' Privacy Issue," Jan. 13, 2011, http://news.cnet.com/8301-30685_3-20028397-264.html.

52   Lance Whitney, "IE Users Can Now Delete Flash Cookies," May 4, 2011, http://news.cnet.com/8301-10805_3-20059653-75.html.

53   Ibid.

54   U.S. Patent Application No. US2010/0010993 A1 (filed March 31, 2009, published Jan 14, 2010).

55   Julia Angwin and Emily Steel, "Web's Hot New Commodity: Privacy," *The Wall Street Journal*, Feb. 28, 2011, at A1, http://online.wsj.com/article/SB10001424052748703529004576160764037920274.html.

56   Rob Frappier, "Changing Our Name, but Not Our Mission," Jan. 12, 2011, www.reputation.com/blog/2011/01/12/changing-our-name-but-not-our-mission/.

57   www.internetreputationmanagement.com; www.reputationhawk.com; http://emarketing.netsmartz.net/internetlaw_online_reputation_management.asp; www.reputationdr.com.

58   Press Release, "Reputation.com Honored as World Economic Forum Technology Pioneer 2011," Sep. 1. 2010, www.reputation.com/press_room/reputationdefender-honored-as-world-economic-forum-technology-pioneer-2011/.

59   "Frequently Asked Questions," www.reputation.com/faq.

60   Ibid.; David Silverberg, "Companies Destroy Data and Tweak Google Search Results to Repair Online Reputations," July 23, 2007, www.digitaljournal.com/article/209811/Companies_Destroy_Data_and_Tweak_Google_Search_Results_to_Repair_Online_Reputations.

61   Maureen Callahan, "Untangling a Web of Lies," Feb. 16, 2007, www.nypost.com/p/entertainment/item_k6T4zNOT1FFDdWjmWVOz8L;jsessionid=84CE063350F7BA1C820E51F55434CEF2.

62   Victoria Murphy Barret, "Anonymity & The Net," Oct. 15, 2007, www.forbes.com/forbes/2007/1015/074.html.

63   Silverberg, "Companies Destroy Data and Tweak Google Search Results to Repair Online Reputations"; Jessica Bennett, "A Tragedy That Won't Fade Away," April 25, 2009, www.newsweek .com/2009/04/24/a-tragedy-that-won-t-fade-away.html.

64   Callahan, "Untangling a Web of Lies."

65   "Frequently Asked Questions," www.reputation.com/faq.

66   David Lazarus, "Forget Privacy; He Sells Your Data," *Los Angeles Times*, June 8, 2010, at A1, published online as "You Won't Find Spokeo Founder Included on His 'People Search' Site," http:// articles.latimes.com/2010/jun/08/business/la-fiw-lazarus-20100608.

67   *In re DoubleClick Inc. Privacy Litigation*, 154 F. Supp. 2d 497, 520 (S.D.N.Y. 2001); *Chance v. Ave. A, Inc.*, 165 F. Supp. 2d 1153, 1158 (W.D. Wash. 2001).

68   18 U.S.C. § 1030.

69   18 U.S.C. §§ 2701-11.

70   *In re DoubleClick Inc. Privacy Litigation*, 154 F. Supp. 2d 497, 510 (S.D.N.Y. 2001).

71   18 U.S.C. §§ 2510-22.

72   *In re DoubleClick Inc. Privacy Litigation*, 154 F. Supp. 2d 497, 518 n. 26 (S.D.N.Y. 2001), *citing Berger v. Cable News Network, Inc.*, 1996 WL 390528 at 3 (D. Mont. 1996) ("[§ 2511(2)(d)] does not apply because this Court does not find that defendants made the recordings for the purpose of committing a crime or tortious act. Instead, the recordings were made for the purpose of producing a news story and for the defendants' commercial gain."), aff'd in part, rev'd in part, 129 F.3d 505 (9th Cir. 1997), vacated and remanded, 526 U.S. 808 (1999), aff'd in relevant part, 188 F.3d 1155 (9th Cir. 1999); *see also, Russell v. ABC, Inc.*, 1995 WL 330920 at 1 (N.D.Ill. 1995) *citing Desnick v. ABC, Inc.*, 44 F.3d 1345, 1353-54 (7th Cir. 1995).

73   *People v. Klapper*, 902 N.Y.S.2d 305 (N.Y. Crim. 2010).

74   Ibid.

75   *Miller v. Meyers*, 766 F Supp. 2d 919 (W.D. Ark. 2011).

76   *O'Brien v. O'Brien*, 899 So. 2d 1133 (Fla. Dist. App. 2005).

77   Jessica Belskis, "Applying the Wiretap Act to Online Communications after *United States v. Councilman*," 2 *Shidler Journal of Law, Commerce and Technology* 18, 24 (2006).

78   *Valentine v. NebuAd*, 2011 WL 1296111 (N.D. Cal. 2011).

79   As of the time this book went to the press, the parties were engaged in settlement negotiations. Proposed Settlement and Release, *Valentine v. NebuAd*, C08-051113 THE (N.D. Cal. Aug. 16, 2011).

80   Wendy Davis, "Case Closed: NebuAd Shuts Down," Online Media Daily, May 18, 2009, www .mediapost.com/publications/?fa=Articles.showArticle&art_aid=106277; Steve Stecklow and Paul Sonne, "Shunned Profiling Technology on the Verge of Comeback," Nov. 24, 2010, http://online.wsj .com/article/SB10001424052748704243904575630751094784516.html; Letter to Hon. Donald S. Clark from J. Brooks Dobbs, "Phorm Inc's Comments on Protecting Consumer Privacy in an Era of Rapid Change," Feb. 18, 2011, www.ftc.gov/os/comments/privacyreportframework/00353-57888.pdf; Jeremy Kirk, "Kindsight Meshes Security Service with Targeted Ads," Nov. 15, 2010, www.pcworld .com/businesscenter/article/210647/kindsight_meshes_security_service_with_targeted_ads.html.

81   *LaCourt v. Specific Media*, 2011 WL 1661532 (C.D. Cal. 2011).

82   Matthew Lasar, "Google, Facebook: 'Do Not Track' Bill a Threat to California Economy," May 6, 2011, http://arstechnica.com/tech-policy/news/2011/05/google-facebook-fight-california-do-not-track -law.ars.

83   "About the Federal Trade Commission," www.ftc.gov/ftc/about.shtm.

84   15 U.S.C. § 45(a)(1).

85   15 U.S.C. § 53(b); 15 U.S.C. § 57(a); 15 U.S.C. § 56(a); 15 U.S.C. § 57(b).

86   15 U.S.C. § 45(b).

87   16 C.F.R. § 2.34; 16 C.F.R. § 325; 16 C.F.R § 2.31.

88   15 U.S.C. § 45(b); 16 C.F.R. § 3.1; 16 C.F.R. § 3.41.

89   Joshua Gomez, Travis Pinnick, and Ashkan Soltani, KnowPrivacy.org Report, U.C. Berkeley School of Information, June 1, 2009, at 18, http://knowprivacy.org/report/KnowPrivacy_Final_Report.pdf.

90   Complaint at 3, *Federal Trade Commission v. Echometrix, Inc.*, No. CV-10-5516 (E.D.N.Y. Nov. 30, 2010).

91   Ibid.

92   Ibid.

93   Stipulated Final Order for Permanent Injunction and Other Equitable Relief, *Federal Trade Commission v. Echometrix, Inc.*, No. CV-10-5516 (E.D.N.Y. Nov. 30, 2010).

94   Louise Story, "F.T.C. Takes a Look at Web Marketing," *The New York Times*, Nov. 2, 2007, at C8, published online as "F.T.C. Member Vows Tighter Controls of Online Ads," www.nytimes.com/2007/11/02/technology/02adco.html.

4. Technology and Fundamental Rights

1   Alfred Lief, *Brandeis: The Personal History of an American Ideal* (Freeport, N.Y.: Books for Libraries Press, 1971; first published 1936), 51.

2   James H. Barron, "Warren and Brandeis, The Right to Privacy, 4 *Harv. L. Rev.* 193 (1890): Demystifying a Landmark Citation," 13 *Suffolk University Law Review* 875 (1979).

3   "Wanamakers," *The Lancaster Daily Intelligencer*, Oct. 26, 1888 (advertisement); "A Photographic Novelty," *New York Daily Tribune*, Dec. 12, 1888.

4   "The Kodak Fiend," *Hawaiian Gazette*, Dec. 9, 1890, at 5.

5   Godkin, "The Rights of the Citizen, IV.-To His Own Reputation," 8 *Scribner's Magazine* 58 (July 1890); Barron, "Warren and Brandeis, The Right to Privacy."

6   Samuel D. Warren and Louis D. Brandeis, "The Right to Privacy," 4 *Harvard Law Review* 193 (1890).

7   *Olmstead v. U.S.*, 277 U.S. 438 (1928).

8   *Kyllo v. U.S.*, 533 U.S. 27 (2001).

9   *Kyllo v. U.S.*, 533 U.S. 27, 31 (2001).

10   *Cruzan v. Dir., Missouri Dept. of Health*, 497 U.S. 261, 356 (1990) (Stevens, J., dissenting).

11   *Norman-Bloodsaw v. Lawrence Berkeley Laboratory*, 135 F.3d 1260, 1269 (9th Cir. 1998).

12   Genetic Information Nondiscrimination Act, Pub. L. No. 110-233, 122 Stat. 881 (codified in scattered sections of U.S.C. titles 26, 29, and 42).

13   Michael Dolan, "The Bork Tapes," www.theamericanporch.com/bork5.htm, originally published in *Washington City Paper*, Sep. 25–Oct. 1, 1987.

14   S. Report No. 100-599, at 5-6 (1988), reprinted in 1988 U.S.C.C.A.N. 4342-1, 4342-5, 4342-6.

15   Ibid. at 6-7 (1988), reprinted in 1988 U.S.C.C.A.N. 4342-1, 4342-6, 4342-7.

16   Ibid. at 7 (1988), reprinted in 1988 U.S.C.C.A.N. 4342-1, 4342-7.

17   Ibid.

18   "About Our Community," City of Coalinga, www.coalinga.com/?pg=1.

19   *Moreno v. Hartford Sentinel, Inc.*, 172 Cal App. 4th 1125 (Cal. Ct. App. 2009).

20   Appellant's Opening Brief at 3-4, *Moreno v. Hanford Sentinel, Inc.*, No. F054138 (Cal. Ct. App. Aug. 2, 2010).

21 Ibid.

22 "Jury Correct in MySpace Ruling, Fresno County Case Had Far-Reaching First Amendment Implications," editorial, *The Fresno Bee*, Sep. 23, 2010, at B4; Pablo Lopez, "Coalinga Grad Sues Principal over MySpace Rant," Sep. 19, 2010, www.fresnobee.com/2010/09/19/2084251/jury-to-decide-on-coalinga-grads.html.

23 Linda Holmes, "Your Privacy Rights on MySpace and Facebook May Be Less than You Think," April 8, 2010, http://public.getlegal.com/articles/online-privacy.

24 *Moreno v. Hartford Sentinel, Inc.*, 172 Cal App. 4th 1125 (Cal. Ct. App. 2009).

25 John Ellis, "Coalinga Grad Loses MySpace Lawsuit," *The Fresno Bee*, Sep. 20, 2010, http://0166244 .blogspot.com/2010/09/heres-your.html.

26 *People v. Klapper*, 902 N.Y.S.2d 305 (N.Y. Crim. Ct. 2010), addressing whether the installation and use of key-logging software to access an employee's password-protected emails violated New York's statute against unauthorized use of a computer.

27 S. Report No. 100-599, at 7 (1988), reprinted in 1988 U.S.C.C.A.N. 4342-1, 4342-7.

28 *Olmstead v. United States*, 277 U.S. 438, 472-473 (1928) (Brandeis, J., dissenting).

29 Eli Pariser, *The Filter Bubble: What the Internet Is Hiding from You* (New York: Penguin Press, 2011), 177–178.

30 Ibid.

5. The Right to Connect

1   E. B. Boyd, "How Social Media Accelerated the Uprising in Egypt," Jan. 31, 2011, www.fastcompany .com/1722492/how-social-media-accelerated-the-uprising-in-egypt.

2   Post by egypt69, "Egypt's 25th of January Revolution," Jan. 23, 2011, www.skyscrapercity.com/archive /index.php/t-1307931.html (linking to www.facebook.com/event.php?eid=115372325200575, though the event page has since been removed).

3   Will Heaven, "Egypt and Facebook: Time to Update Its Status," *NATO Review*, 2011, www.nato.int /docu/review/2011/Social_Medias/Egypt_Facebook/EN/index.htm.

4   Brian Ross and Matthew Cole, "Egypt: The Face That Launched a Revolution," Feb. 4, 2011, http:// abcnews.go.com/Blotter/egypt-face-launched-revolution/story?id=12841488&page=1.

5   Ernesto Londono, "Egyptian Man's Death Became Symbol of Callous State," Feb. 9, 2011, www .washingtonpost.com/wp-dyn/content/article/2011/02/08/AR2011020806360.html.

6   Heaven, "Egypt and Facebook: Time to Update Its Status."

7   Jim Michaels, "Tech-savvy Youths Led the Way in Egypt Protests," Feb. 7, 2011, www.usatoday.com /news/world/2011-02-07-egyptyouth07_ST_N.htm.

8   Erik Schonfeld, "The Egyptian Behind #Jan25: 'Twitter Is a Very Important Tool for Protesters,'" Feb. 16, 2011, http://techcrunch.com/2011/02/16/jan25-twitter-egypt/ (linking to http://twitter.com/#! /alya1989262/status/26353718601449472).

9   Iyad El-Baghdadi, "Meet Asmaa Mahfouz and the Vlog that Helped Spark the Revolution," Feb. 1, 2011, www.youtube.com/watch?v=SgjIgMdsEuk.

10 Emad Mekay, "Arab Women Lead the Charge," Feb. 11, 2011, http://ipsnews.net/news.asp ?idnews=54439.

11 Aya A. Khalil, "Thousands Fill the Streets in Egypt Protests (updates)," Feb. 26, 2011, http://ayakhalil .blogspot.com/2011/02/thousands-fill-streets-in-egypt.html.

12 Noah Shachtman, "How Many People Are in Tahrir Square? Here's How to Tell [Updated]," Feb. 1, 2011, www.wired.com/dangerroom/2011/02/how-many-people-are-in-tahrir-square-heres-how-to-tell/.

13 Michaels, "Tech-savvy Youths Led the Way in Egypt Protests."

14   James Cowie, Chief Technology Officer of Renesys, "Libya Disconnect," Feb. 18, 2011, updated March 4, 2011, www.renesys.com/blog/2011/02/libyan-disconnect-1.shtml.

15   Bill Woodcock, "Overview of the Egyptian Internet Shutdown," Packet Clearing House Presentation, Feb. 2011, www.pch.net/resources/misc/Egypt-PCH-Overview.pdf.

16   Ryan Singel, "Egypt Shut Down Net with Big Switch, Not Phone Calls," Feb. 10, 2011, www.wired .com/threatlevel/2011/02/egypt-off-switch/.

17   Cowie, "Libya Disconnect."

18   Ibid.

19   Michaels, "Tech-savvy Youths Led the Way in Egypt Protests."

20   Diane Macedo, "Egyptians Use Low-tech Gadgets to Get Around Communications Block," Jan. 28, 2011, www.foxnews.com/scitech/2011/01/28/old-technology-helps-egyptians-communications -black/.

21   "Modern Protests in Ancient Egypt," Feb. 3, 2011, http://secondedition.wordpress.com/2011/02/03 /modern-protests-in-ancient-egypt/.

22   Wagner James Au,"Egyptians Worldwide Gather in Second Life to Share Resources, Information, Support for Uprising," Jan. 30, 2011, http://nwn.blogs.com/nwn/2011/01/egyptians-in-second-life .html.

23   Cowie, "Libya Disconnect."

24   "The Economic Impact of Shutting Down Internet and Mobile Phone Services in Egypt," Feb. 4, 2011, www.oecd.org/document/19/0,3746,en_2649_33703_47056659_1_1_1_1,00.html.

25   Richard Hartley-Parkinson, "Meet My Daughter 'Facebook': How One New Egyptian Father Is Commemorating the Part the Social Network Played in Revolution," Feb. 21, 2011, www.dailymail .co.uk/news/article-1358876/Baby-named-Facebook-honour-social-network-Egypts-revolution.html.

26   Alexia Tsotsis, "To Celebrate the #Jan25 Revolution, Egyptian Names His Firstborn 'Facebook,'" Feb. 19, 2011, http://techcrunch.com/2011/02/19/facebook-egypt-newborn/ (translated from Al-Ahram, Feb. 18, 2011, at 2, www.ahram.org.eg/pdf/Zoom_1500/Index.aspx?ID=45364).

27   Jesse Emspak, "Libya Blocks Internet Traffic," International Business Times, March 4, 2011, www .ibtimes.com/articles/118969/20110304/libya-cuts-off-internet-engages-kill-switch.htm.

28   Cowie, "Libya Disconnect."

29   Emspak, "Libya Blocks Internet Traffic."

30   Economist Intelligence Unit Ltd., "Libya: Privatisation Possibilities," March 19, 2007, http:// globaltechforum.eiu.com/index.asp?layout=rich_story&channelid=4&categoryid=31&title=Libya% 3A+Privatisation+possibilities&doc_id=10336.

31   Emspak, "Libya Blocks Internet Traffic."

32   Cybersecurity and Internet Freedom Act of 2011, S. 413 §§ 101, 249(a)(3)(A), 112th Cong., 1st Session (2011).

33   See, "Internet Filtering in Egypt," OpenNet Initiative, Aug. 6, 2009, http://opennet.net/research /profiles/egypt; "Internet Filtering in Libya," OpenNet Initiative, Aug. 6, 2009, http://opennet.net /research/profiles/libya.

34   Jon Swartz, "'Kill Switch' Internet Bill Alarms Privacy Experts," Feb. 21, 2011, http://abcnews .go.com/Technology/kill-switch-internet-bill-alarms-privacy-experts/story?id=12922845.

35   Jennifer Valentino-DeVries, "How Egypt Killed the Internet," The Wall Street Journal Digits blog, Jan. 28, 2011, http://blogs.wsj.com/digits/2011/01/28/how-egypt-killed-the-internet/.

36   QuinStreet Inc., "Top 23 U.S. ISPs by Subscriber: Q3 2008," 2008, www.isp-planet.com/research /rankings/usa.html.

37   "Reaching for the Kill Switch," Feb. 10, 2011, www.economist.com/node/18112043.

38   Carolyn Duffy Marsan, "U.S. Plots Major Upgrade to Internet Router Security," Jan. 15, 2009, www
      .networkworld.com/news/2009/011509-bgp.html?page=1; Carolyn Duffy Marsan, "Feds to Shore
      Up Net Security," Jan. 19, 2009, www.pcworld.com/businesscenter/article/157909/feds_to_shore_up
      _net_security.html.

39   Erik Turnquist, "Government Plans Massive Internet Backbone Security Upgrade," University
      of Washington Computer Security Research and Course Blog, Jan. 16, 2009, https://cubist
      .cs.washington.edu/Security/2009/01/16/current-event-government plans-massive-internet
      -backbone-security-upgrade/.

40   The Associated Press, "Boeing Buying Cybersecurity Firm Narus," July 8, 2010, www.businessweek
      .com/ap/financialnews/D9GQTHC00.htm; Robert Poe, "The Ultimate Net Monitoring Tool," May
      17, 2006, www.wired.com/print/science/discoveries/news/2006/05/70914.

41   John Markoff and Scott Shane, "Documents Show Link Between AT&T and Agency in
      Eavesdropping Case," *The New York Times*, April 13, 2006, at A17, www.nytimes.com/2006/04/13/us
      /nationalspecial3/13nsa.html?T=&n=Top/News/Business/Companies/AT&_r=1&pagewanted=all.

42   Ibid.

43   "EFF's Case Against AT&T," Electronic Frontier Foundation, www.eff.org/nsa/hepting.

44   The Associated Press, "Boeing Buying Cybersecurity Firm Narus."

45   Worldview, "The YouTube Interview with President Obama, 2011," Jan. 27, 2011, www.youtube
      .com/watch?v=etaCRMEFRy8&t=18m31s.

46   U.S. Department of State, Bureau of International Information Programs, "Rights of the People,
      Individual Freedom and the Bill of Rights, Chapter 4, Freedom of the Press," www.4uth.gov.ua/usa
      /english/society/rightsof/press.htm.

47   *McIntyre v. Ohio Elections Commission*, 514 U.S. 334, 341-342 (1995).

48   Ibid. at 342.

49   *Globe Newspaper Co. v. Superior Ct.*, 457 U.S. 596 (1982).

50   *See, e.g., Talley v. California*, 362 U.S. 60, 64 (1960) (finding that a city ordinance that required the
      name of the preparer to be printed on handbills "would tend to restrict the freedom to distribute
      information and thereby freedom of expression").

51   *See, e.g., Doe v. 2The Mart.com*, 140 F. Supp. 2d 1088 (W.D. Wash. 2001) (noting that First
      Amendment protections extended to noncore speech, including calling a corporation a liar, criminal,
      and defrauder).

52   *See, e.g., McIntyre v. Ohio Elections Commission*, 514 U.S. 334, 357 (1995).

53   *See, e.g., Independent Newspapers, Inc. v. Brodie*, 966 A.2d 432, 444 (Md. 2009) (noting that "viable
      causes of actions for defamation should not be barred in the Internet context"); *In re Subpoena
      Duces Tecum to America Online, Inc.*, 2000 WL 1210372 at 6 (Va. Cir. Ct. 2000) (redress should not
      be foreclosed because a wrongdoer hides behind "an illusory shield of purported First Amendment
      rights").

54   Jon Orwant, "Find Out What's in a Word, or Five, with the Google Books Ngram Viewer," Inside
      Google Books blog, Dec. 16, 2010, http://booksearch.blogspot.com/2010/12/find-out-whats-in-word
      -or-five-with.html.

55   Leonid Taycher, "Books of the world, stand up and be counted! All 129,864,880 of you," Inside
      Google Books blog, Aug. 5, 2010, http://booksearch.blogspot.com/2010/08/books-of-world-stand-up
      -and-be-counted.html.

56   Bill Rounds, "How to Make Anonymous Comments on a Website," Jan. 2, 2011, www.howtovanish
      .com/2011/01/how-to-make-anonymous-comments-on-a-website/.

57   "How to Blog Safely (About Work or Anything Else)," Electronic Frontier Foundation, April 6, 2005,
      updated May 31, 2005, www.eff.org/wp/blog-safely.

58   "Tor: Overview," www.torproject.org/about/overview.html.en.

59   Hillary Rodham Clinton, "Internet Rights and Wrongs: Choices & Challenges in a Networked World," speech, Feb. 15, 2011, www.state.gov/secretary/rm/2011/02/156619.htm.

60   *Reno v. Am. Civ. Liberties Union*, 521 U.S. 844, 870 (1997).

61   *Doe v. 2TheMart.com*, 140 F. Supp. 2d 1088, 1091-1092 (W.D. Wash. 2001).

62   *Doe v. 2TheMart.com*, 140 F. Supp. 2d 1088, 1090 (W.D. Wash. 2001).

63   *Dendrite Int'l, Inc. v. Doe No. 3*, 775 A.2d 756, 767 (N.J. Super. App. Div. 2001).

64   U.N. Special Rapporteur, "Promotion and Protection of the Right to Freedom of Opinion and Expression: Rep. of the Special Rapporteur," ¶ 82, U.N. Doc. A/HRC/17/27 May 16, 2011, www2 .ohchr.org/english/bodies/hrcouncil/docs/17session/A.HRC.17.27_en.pdf.

65   Conseil Constitutionnel [Constitutional Court] decision No. 2009-580, June 10, 2009 (Fr.), 1, 4 (quoting Article 11 of the Declaration of the Rights of Man and the Citizen of 1789) (translation www.conseil-constitutionnel.fr/conseil-constitutionnel/root/bank_mm/anglais/2009_580dc.pdf).

66   Eesti Vabariigi Põhiseadus [Constitution of the Republic of Estonia], June 28, 1999, art. 44-45 (Est.) (official English translation www.president.ee/en/republic-of-estonia/the-constitution/index.html).

67   Telekommunikatsiooniseadus [Telecommunications Act] (Est.) (translation www.legaltext.ee/text/en /X30063K6.htm).

68   Jenny Wittauer, "Tiger Leap Project in Estonia: Free Internet Access," School of Journalism, Utrecht, The Netherlands, June 18, 2010, www.schoolvoorjournalistiek.com/europeanculture09/?p=1742.

69   "World Press Freedom Index 2010 The Rankings," Reporters Without Borders, Oct. 10, 2010, www .rsf.org/IMG/CLASSEMENT_2011/GB/C_GENERAL_GB.pdf

70   The Constitution of the Arab Republic of Egypt, art. 48, translation www.egypt.gov.eg/english/laws /constitution/default.aspx.

71   U.S. Department of State, Bureau of Democracy, Human Rights, and Labor, "Human Rights Report: Egypt," www.state.gov/g/drl/rls/hrrpt/2009/nea/136067.htm.

72   Ibid.

73   Frank La Rue, "Report of the Special Rapporteur on the Promotion and Protection of the Right to Freedom of Opinion and Expression," ¶ 2, U.N. Doc. A/HRC/17/27 May 16, 2011, www2.ohchr.org /english/bodies/hrcouncil/docs/17session/A.HRC.17.27_en.pdf.

74   Ibid. at ¶ 82.

75   G8 Declaration: Renewed Commitment for Freedom and Democracy, G8 Summit of Deauville, France, May 26–27, 2011, www.g20-g8.com/g8-g20/g8/english/live/news/renewed-commitment-for -freedom-and-democracy.1314.html.

76   *New York Times Co. v. Sullivan*, 376 U.S. 254 (1964).

77   Morris D. Forkosch, "Freedom of the Press: Croswell's Case," 33 *Fordham Law Review* 415, 417, 429 (1964–1965).

78   Ron Chernow, *Alexander Hamilton* (New York: Penguin, 2004), 669.

79   Mike Farrell and Mary Carmen Cupito, *Newspapers: A Complete Guide to the Industry* (New York: Peter Lang, 2010), 26.

80   Edward Moyer, "Stuxnet Worm Strike Iranian Nuclear Plant," Sep. 27, 2011, www.zdnet.com/news /stuxnet-worm-strike-iranian-nuclear-plant/468939.

81   Jonathan Fildes, "Stuxnet Virus Targets and Spread Revealed," Feb. 15, 2011, www.bbc.co.uk/news /technology-12465688.

82   William J. Broad, "Israeli Test on Worm Called Crucial in Iran Nuclear Delay," *The New York Times*, Jan. 16, 2011, at A1, www.nytimes.com/2011/01/16/world/middleeast/16stuxnet.html.

83   Fildes, "Stuxnet Virus Targets and Spread Revealed."

84   Jonathan Zittrain, "Will the U.S. Get an Internet 'Kill Switch'?," March 4, 2011, www.technologyreview
      .com/web/32451/page1/.

85   Before the House Committee on Foreign Affairs, "Recent Developments in Egypt and Lebanon:
      Implications for U.S. Policy and Allies in the Broader Middle East," Statement of Congressman Chris
      Smith, (Feb. 10, 2011), http://foreignaffairs.house.gov/112/64483.pdf.

86   "Recent Developments in Egypt and Lebanon: Implications for U.S. Policy and Allies in the Broader
      Middle East," Statement of Congressman William Keating Before the House Committee on Foreign
      Affairs, Feb. 10, 2011, http://foreignaffairs.house.gov/112/64483.pdf.

87   Office of Congressman William Keating Press Release, "Social Media Is Being Used as a Weapon,"
      Feb. 10, 2011, http://keating.house.gov/index.php?option=com_content&view=article&id=86:keat
      ing-social-media-is-being-used-as-a-weapon&catid=1:press-releases&Itemid=13.

88   Ibid.

89   "An EUM Bellwether? India/US Arms Deals Face Crunch Over Conditions," Nov. 9, 2010, www
      .defenseindustrydaily.com/IndiaUS-Arms-Deals-Facing-Crunch-Over-Conditions-05285/.

90   Charles Bremner, "Top French Court Rips Heart Out of Sarkozy Internet Law," June 11, 2009, http://
      technology.timesonline.co.uk/tol/news/tech_and_web/article6478542.ece.

91   Conseil Constitutionnel [Constitutional Court] decision No. 2009-580, June 10, 2009, Fr., 1, 4,
      quoting Article 11 of the Declaration of the Rights of Man and the Citizen of 1789, translation www
      .conseil-constitutionnel.fr/conseil-constitutionnel/root/bank_mm/anglais/2009_580dc.pdf.

92   http://doctoredreviews.com.

93   Thomas Jefferson, "Letter to Edward Carrington," Jan. 16, 1787, in Thomas Jefferson: Writings, ed.
      Merrill D. Peterson (New York: Library of America, 1984), 880.

94   U.N. Special Rapporteur, "Promotion and Protection of the Right to Freedom of Opinion and
      Expression: Rep. Of the Special Rapporteur," U.N. Doc. A/HRC/17/27, May 16, 2011, at ¶ 82,
      www2.ohchr.org/english/bodies/hrcouncil/docs/17session/A.HRC.17.27_en.pdf.

95   Lyrissa Barnett Lidsky, "Silencing John Doe: Defamation and Discourse in Cyberspace," 49 Duke
      Law Journal 855, 860 (2000).

96   Ibid.

97   Bianca Bosker, "Facebook's Randi Zuckerberg: Anonymity Online 'Has to Go Away,'" July 27, 2011,
      www.huffingtonpost.com/2011/07/27/randi-zuckerberg-anonymity-online_n_910892.html?icid=
      maing-grid7|main5|dl6|sec3_lnk3|81656.

98   Marshall Kirkpatrick, "Google, Privacy and the Explosion of New Data," Aug. 4, 2010, http://
      techonomy.typepad.com/blog/2010/08/google-privacy-and-the-new-explosion-of-data.html.

99   Viktor Mayer-Schönberger, Delete: The Virtue of Forgetting in the Digital Age (Princeton, NJ:
      Princeton University Press, 2009), 141, citing William Seltzer and Margo Anderson, "The Dark Side
      of Numbers: The Role of Population Data Systems in Human Rights Abuses," 68 Social Research
      481, 486 (2001).

6. Freedom of Speech

1   Emmett v. Kent School District No. 415, 92 F. Supp. 2d 1088 (W.D. Wash. 2000).

2   "SC Firefighter-Paramedic Fired over Facebook Video Post," Feb. 26, 2010, www.ems1.com
     /ems-management/articles/765102-SC-firefighter-paramedic-fired-over-Facebook-video-post/.

3   Colleton County Fire-Rescue Termination of Employment Memorandum, Feb. 11, 2010, http://
     wcsc.images.worldnow.com/images/incoming/pdf/termination.pdf.

4   A.B. v. State, 885 N.E.2d 1223 (Ind. 2008).

5   "After ACLU of Colorado Intervention, High School Student Suspended for Off-Campus Internet Posting Is Back in School," Feb. 21, 2006, www.aclu.org/free-speech/after-aclu-colorado-intervention -high-school-student-suspended-campus-internet-posting-b.

6   Missouri, for example.

7   *Beussink v. Woodland R-IV School District*, 30 F. Supp. 2d 1175 (E.D. Mo. 1998).

8   *Layshock v. Hermitage School District*, 593 F.3d 249 (3rd Cir. 2010), aff'd on reh'g en banc, 2011 WL 2305970 (3rd Cir. 2011).

9   Mary Helen Miller, "East Stroudsburg U. Suspends Professor for Facebook Posts," Feb. 26, 2010, http://chronicle.com/blogPost/East-Stroudsburg-U-Suspends/21498.

10 Raegan Medgie, "ESU Professor Back in Class," March 31, 2010, www.wnep.com/news/county bycounty/wnep-mon-esu-professor-back-in-class,0,2905966.story.

11 Dalia Fahmy, "Professor Suspended After Joke About Killing Students on Facebook," March 3, 2010, http://abcnews.go.com/Business/PersonalFinance/facebook-firings-employees-online-vents-twitter -postings-cost/story?id=9986796.

12 Miller, "East Stroudsburg U. Suspends Professor for Facebook Posts."

13 Medgie, "ESU Professor Back in Class."

14 Miller, "East Stroudsburg U. Suspends Professor for Facebook Posts."

15 *Pietrylo v. Hillstone Restaurant Group*, 2008 WL 6085437 (D. N.J. 2008).

16 Ibid.

17 *J.S. v. Bethlehem Area School District*, 757 A.2d 412 (Pa. Commw. Ct. 2000).

18 *J.S. v. Blue Mountain School District*, 593 F.3d 286 (3rd Cir. 2010), aff'd in part, rev'd in part and remanded on reh'g en banc (3rd Cir. 2011).

19 Associated Press, "Federal Judge Halts Suspension of Student Punished over Web Site," Feb. 24, 2000, www.firstamendmentonline.net/%5Cnews.aspx?id=7559.

20 *Emmett v. Kent School District No. 415*, 92 F. Supp. 2d 1088, 1090 (W.D. Wash. 2000).

21 *Tinker v. Des Moines Independent Community School District*, 393 U.S. 503, 509 (1969) (invalidating the school's ban on wearing black armbands protesting the Vietnam War).

22 *Emmett v. Kent School District No. 415*, 92 F. Supp. 2d 1088, 1090 (W.D. Wash. 2000), *citing Tinker v. Des Moines Independent Community School District*, 393 U.S. 503, 509 (1969).

23 *Emmett v. Kent School District No. 415*, 92 F. Supp. 2d 1088, 1090 (W.D. Wash. 2000).

24 Ibid.

25 *Beussink v. Woodland R-IV School District*, 30 F. Supp. 2d 1175, 1180-1182 (E.D. Mo. 1998), *citing Terminiello v. City of Chicago*, 337 U.S. 1, 4 (1949).

26 *Beussink v. Woodland R-IV School District*, 30 F. Supp. 2d 1175, 1182 (E.D. Mo. 1998).

27 *A.B. v. State*, 885 N.E.2d 1223, 1227 (Ind. 2008).

28 *J.S. v. Bethlehem Area School District*, 807 A.2d 847, 865-869 (Pa. 2002).

29 *J.S. v. Bethlehem Area School District*, 757 A.2d 412, 428 n.6 (Pa. Commw. Ct. 2000).

30 *Killion v. Franklin Regional School District*, 136 F. Supp. 2d 446 (W.D. Pa. 2001).

31 *J.S. v. Blue Mountain School District*, 593 F.3d 286 (3rd Cir. 2010), aff'd in part, rev'd in part and remanded on reh'g en banc (3rd Cir. 2011).

32 Kaitlin Madden, "Ten Tweets That Could Get You Fired," April 6, 2011, www.theworkbuzz.com /careers/tweets-could-get-you-fired/.

33 David Kirkpatrick, *The Facebook Effect* (New York: Simon & Schuster, 2010), 211.

34   Andrew Levy, "Teenage Office Worker Sacked for Moaning on Facebook About Her 'Totally Boring' Job," Feb. 26, 2009, www.dailymail.co.uk/news/article-1155971/Teenage-office-worker-sacked-moaning-Facebook-totally-boring-job.html.

35   Kirkpatrick, *The Facebook Effect*, 211.

36   Kaitlin Madden, "12 Ways to Get Fired for Facebook," Aug. 9, 2010, www.careerbuilder.ca/Article/CB-606-Workplace-Issues-12-Ways-to-Get-Fired-for-Facebook/.

37   Wilma B. Liebman, "Decline and Disenchantment: Reflections on the Aging of the National Labor Relations Board," 28 *Berkeley Journal of Employment and Labor Law* 569, 576 (2007) (footnotes omitted).

38   29 U.S.C. §157.

39   Hatzel Vela, "Colleton County Rescue Worker Loses Job over Facebook Post," March 18, 2010, www.live5news.com/story/12047151/colleton-county-rescue-worker-loses-job-over-facebook-post?redirected=true.

40   Declan McCullagh, "Yes, Insults on Facebook Can Still Get You Fired," Nov. 9, 2010, http://news.cnet.com/8301-13578_3-20022276-38.html#ixzz14tyiCByQ.

41   Steven Greenhouse, "Company Accused of Firing over Facebook Post," *The New York Times*, Nov. 8, 2010, at B1, www.nytimes.com/2010/11/09/business/09facebook.html.

42   Ibid.

43   Ibid.

44   Gordon MacMillan, "BBC and Reuters Turn to Social Media Guidelines After Staff Tweet Trouble," April 7, 2011, http://wallblog.co.uk/2011/04/07/bbc-and-reuters-turn-to-social-media-guidelines-after-staff-tweet-trouble/.

45   Steven Greenhouse, "Labor Panel to Press Reuters over Reaction to Twitter Post," *The New York Times*, April 7, 2011, at B3, www.nytimes.com/2011/04/07/business/media/07twitter.html.

46   *Pietrylo v. Hillstone Restaurant Group*, 2008 WL 6085437 (D. N.J. 2008).

47   *Pietrylo v. Hillstone Restaurant Group*, 2009 WL 3128420 (D. N.J. 2009).

48   Bob Cohn, "Dismissed Pierogi Returns to Run Again at PNC Park," June 23, 2010, www.pittsburghlive.com/x/pittsburghtrib/sports/s_687194.html.

49   *Schenck v. United States*, 249 U.S. 47, 52 (1919).

50   "Israeli Military 'Unfriends' Soldier After Facebook Leak," March 4, 2010, http://news.bbc.co.uk/2/hi/8549099.stm.

51   Andie531, "'We Are Cleaning Out a West Bank Village' Facebook Entry Gets Soldier Booted," The Truth Will Set You Free blog, March 4, 2010, http://wakeupfromyourslumber.com/blog/andie531/we-are-cleaning-out-west-bank-village-facebook-entry-gets-soldier-booted.

52   "Quote of the Day," March 3, 2010, www.time.com/time/quotes/0,26174,1969477,00.html.

53   Haaretz Service and Reuters, "IDF Calls Off West Bank Raid Due to Facebook Leak," March 3, 2010, www.haaretz.com/news/idf-calls-off-west-bank-raid-due-to-facebook-leak-1.264065.

54   "Israeli Military 'Unfriends' Soldier After Facebook Leak."

55   *Hammonds v. Aetna Casualty & Surety Co.*, 243 F. Supp. 793, 801 (N.D. Ohio 1965).

56   Madden, "12 Ways to Get Fired for Facebook."

57   Chelsea Conaboy, "For Doctors, Social Media a Tricky Case," *The Boston Globe*, April 20, 2011, at 1, www.boston.com/lifestyle/health/articles/2011/04/20/for_doctors_social_media_a_tricky_case/?page=full.

58   Ibid.; Associated Press, "Doctor Busted for Patient Info Spill on Facebook," April 18, 2011, www.msnbc.msn.com/id/42652527/ns/technology_and_science-security/t/doctor-busted-patient-info-spill-facebook/.

59   Complaint, *In re Kristine Ann Peshek*, Hearing Board of the Illinois Attorney Registration and Disciplinary Commission (Aug. 25, 2009), www.iardc.org/09CH0089CM.html.

60   Sarah Bruce, "Nurse Suspended for Op Pics on Facebook," E-Health Insider, Jan. 25, 2010, www.ehi .co.uk/news/ehi/5579.

61   Emil Protalinski, "Parents Suing Facebook over Photo of Murdered Daughter," March 29, 2011, www.zdnet.com/blog/facebook/parents-suing-facebook-over-photo-of-murdered -daughter/1024?tag=mantle_skin;content.

62   Associated Press/CBS New York, "Ex-NYC EMT Admits Posting Corpse Photo on Facebook," Dec. 10, 2010, http://newyork.cbslocal.com/2010/12/10/ex-nyc-emt-admits-posting-corpse-photo-on -facebook/.

63   *Tatro v. University of Minnesota*, Case No. A10-1440, 2011 WL 2672220 at 9 (Minn. App. 2011).

64   Annie Karni, "Web Site Exposes Bad Tippers in Brooklyn," May 8, 2011, www.nypost.com/p/news /local/brooklyn/keep_the_bleepin_change_4jDvBIHLSv7jsMuG4cIYYL; Adrian Chen, "Brooklyn Delivery Guy Starts Blog Shaming Bad Tippers," April 29, 2011, http://blog.gawker.com/5797195 /brooklyn-delivery-guy-starts-blog-shaming-bad-tippers.

65   *Brents v. Morgan*, 299 S.W. 967 (Ky. Ct. App. 1927).

66   Caitlin Fitzsimmons, "Arkansas School Board Member Resigns over Anti-Gay Facebook Comments," Oct. 30, 2010, www.allfacebook.com/arkansas-school-board-2010-10.

67   Ibid.

68   Cynthia Bowers, "In Wisconsin, Thousands Protest Budget Proposal," Feb. 20, 2011, www.cbsnews .com/stories/2011/02/20/sunday/main20034127.shtml.

69   Adam Weinstein, "Indiana Official: 'Use Live Ammunition' Against Wisconsin Protesters," Feb. 23, 2011, http://motherjones.com/politics/2011/02/indiana-official-jeff-cox-live-ammunition-against -wisconsin-protesters.

70   Ibid.

71   Ibid.

72   International Association of Chiefs of Police Center for Social Media, September 2010 survey on law enforcement's use of social media, at 11, www.iacpsocialmedia.org/Portals/1/documents/Survey%20 Results%20Document.pdf.

73   Erica Goode, "Police Lesson: Social Network Tools Have Two Edges," *The New York Times*, April 7, 2011, at A1, www.nytimes.com/2011/04/07/us/07police.html.

74   Magdalena Sharpe, "Officer Lists Job as 'Human Waste Disposal' on Facebook," Feb. 16, 2011, www .kob.com/article/stories/S1976709.shtml.

75   Goode, "Police Lesson: Social Network Tools Have Two Edges."

76   Albuquerque Police Department General Orders, Addition to the Manual, 1-44 Social Media Policy, Rules 1.44.2A and B, www.scribd.com/doc/50958630/Social-Network-Policy-SOP-Sanctions.

77   Jeff Proctor, "Cop in Facebook Scandal Back on Street," *Albuquerque Journal*, May 6, 2011, www.abqjournal.com/cgi-bin/print_it.pl?page=/news/metro/062240306814newsmetro05-06-11 .htm.

78   Goode, "Police Lesson: Social Network Tools Have Two Edges."

79   Jim Meyer, "The Officer Who Posted Too Much on MySpace," *The New York Times*, March 11, 2009, at A24, www.nytimes.com/2009/03/11/nyregion/11about.html.

80   Ibid.

81   Ibid.

82   *Flaherty v. Keystone Oaks School District*, 247 F. Supp. 2d 698, 702 (W.D. Pa. 2003).

83   ACLU Press Release, "PA School Pays $60,000 to Student Who Was Punished for Private Internet Message," Nov. 18, 2002, www.aclu.org/technology-and-liberty/pa-high-school-pays-60000-student -who-was-punished-private-internet-message.

84   *Flaherty v. Keystone Oaks School District*, 247 F. Supp. 2d 698, 701 (W.D. Pa. 2003).

85   Ibid.

86   *Endicott Interconnect Technologies, Inc. v. N.L.R.B.*, 453 F.3d 532 (D.C. Cir. 2006).

87   *Endicott Interconnect Technologies, Inc. v. N.L.R.B.*, 453 F.3d 532, 537 (D.C. Cir. 2006), *citing Jefferson NLRB v. Electrical Workers Local 1229 (Jefferson Standard)*, 346 U.S. 464, 471 (1953).

88   *Endicott Interconnect Technologies, Inc. v. N.L.R.B.*, 453 F.3d 532, 538 (D.C. Cir. 2006).

89   Rob Quinn, "Domino's Workers Fired for Gross-Out Video," April 15, 2009, www.newser.com /story/56201/dominos-workers-fired-for-gross-out-video.html.

90   Stephanie Clifford, "Video Prank at Domino's Taints Brand," *The New York Times*, April 16, 2009, at B1, www.nytimes.com/2009/04/16/business/media/16dominos.html?_r=1&ref=business.

91   Jennifer Lawinski, "Domino's at Center of Disgusting YouTube Video Prank Closes," Sep. 29, 2009, www.slashfood.com/2009/09/29/dominos-at-center-of-disgusting-youtube-video-prank-closes/.

92   Clifford, "Video Prank at Domino's Taints Brand."

93   Ibid.

94   Lawinski, "Domino's at Center of Disgusting YouTube Video Prank Closes."

95   *Pickering v. Board of Education of Township High School District 205*, 391 U.S. 563 (1968).

96   Christopher Jordan, "Social Media at the Workplace in Germany," *Social Networking in the International Workplace*, American Bar Association, April 20, 2011, at 9; Walter Born, "Germany," *Advisory: European Employment Law Update*, Covington & Burling LLP, Feb. 1, 2011, at 3, www.cov.com/files/Publication/718f2a6b-8f57-4f40-a773-f0c971c99f39/Presentation /PublicationAttachment/78f70ce1-0924-44c3-8b22-046e0b382e3a/European%20Employment%20 Law%20Update.pdf.

97   William McGeveran, "Finnish Employers Cannot Google Applicants," Nov. 15, 2006, http://blogs .law.harvard.edu/infolaw/2006/11/15/finnish-employers-cannot-google-applicants/.

7. Lethal Advocacy

1   Justin Piercy, "Missing Student's Family Criticizes Ottawa Police," March 22, 2008, www.thestar .com/article/349669.

2   "Death Online," *The Fifth Estate*, Canadian Broadcasting Corporation, Oct. 9, 2009, www.cbc.ca/ fifth/2009-2010/death_online/; "Nadia's Ottawa 'Hell' Described: Grieving Father Says Daughter's Troubles Were Kept from Him," April 21, 2008, www.metronews.ca/ottawa/local/article/42857.

3   "Missing Teen's Family Offers $50,000 Reward," March 16, 2008, www.canada.com/ottawa citizen/news/story.html?id=286d9ad6-6715-408f-8d16-d73a4f2a4719&k=2760; Michele Mandel, "Parents Losing Hope," *The Ottawa Sun*, April 7, 2008, http://cnews.canoe.ca/CNEWS/Canada /2008/04/07/5216056-sun.html.

4   Bob McKeown, "Dangerous Connection," *Dateline NBC*, Aug. 20, 2010, www.msnbc.msn.com /id/38739087/ns/dateline_nbc/.

5   "Death Online."

6   "Kajouji's Video Diary Shows Path to Suicide," Oct. 9, 2009, www.cbc.ca/news/canada/ottawa /story/2009/10/09/ottawa-kajouji-fifth-estate-diary-suicide.html.

7   "Death Online."

8   "Kajouji's Video Diary Shows Path to Suicide."

9   "Death Online."

10   "Kajouji's Video Diary Shows Path to Suicide."

11   "Death Online."

12   McKeown, "Dangerous Connection."

13   Ibid.

14   Ibid.

15   Statement by Krystal Leonov on "Death Online."

16   "Nadia's Ottawa 'Hell' Described."

17   "Death Online."

18   "Kajouji's Video Diary Shows Path to Suicide.".

19   *Minnesota v. Melchert-Dinkel*, 2011 WL 893506 (Minn. Dist. Ct. 2011).

20   Ibid.

21   Ibid. at Findings of Fact ¶ 31.

22   Ibid.

23   Ibid.

24   Ibid.

25   Summons, *Minnesota v. Melchert-Dinkel*, No. 66-CR-10-1193 (Minn. Dist. Ct. April 23, 2010).

26   McKeown, "Dangerous Connection."

27   Monica Davey, "Online Talk, Suicides and a Thorny Court Case," *The New York Times,* May 14, 2010, at A1.

28   David Brown, "Village Sleuth Unmasks US Internet Predator Behind Suicide 'Pacts,'" March 20, 2010, www.timesonline.co.uk/tol/news/uk/crime/article7069144.ece?token=null&offset=0&page=1.

29   Ibid.

30   *Minnesota v. Melchert-Dinkel*, 2011 WL 893506 (Minn. Dist. Ct. 2010).

31   Brown, "Village Sleuth Unmasks US Internet Predator Behind Suicide 'Pacts.'"

32   Statement by Celia Blay on McKeown, "Dangerous Connection."

33   Ibid.

34   *Minnesota v. Melchert-Dinkel*, 2011 WL 893506 (Minn. Dist. Ct. 2011).

35   McKeown, "Dangerous Connection."

36   "In Search of Hope," *The Ottawa Citizen*, April 13, 2008, www.canada.com/ottawacitizen/news/story.html?id=95ce7f31-d990-4354-a0d6-e3c5558b4847&k=6683.

37   *Minnesota v. Melchert-Dinkel*, 2011 WL 893506 (Minn. Dist. Ct. 2011).

38   Ibid. at Findings of Fact ¶ 31.

39   Ibid.

40   Ibid.

41   Michele Mandel, "Parents Losing Hope," *The Ottawa Sun*, April 7, 2008, http://cnews.canoe.ca/CNEWS/Canada/2008/04/07/5216056-sun.html.

42   *Minnesota v. Melchert-Dinkel*, 2011 WL 893506 (Minn. Dist. Ct. 2011).

43   Mandel, "Parents Losing Hope."

44   *Minnesota v. Melchert-Dinkel*, 2011 WL 893506 at Findings of Fact ¶ 31 (Minn. Dist. Ct. 2011).

45   Ibid.

46   Ibid. at Findings of Fact ¶ 15.

47 McKeown, "Dangerous Connection."

48 *Minnesota v. Melchert-Dinkel*, 2011 WL 893506 at Findings of Fact ¶ 35 (Minn. Dist. Ct. 2011).

49 Ibid. at Findings of Fact ¶ 34.

50 Summons, *Minnesota v. Melchert-Dinkel*, Case No. 66-CR-10-1193 (Minn. Dist. Ct. April 23, 2010).

51 *Minnesota v. Melchert-Dinkel*, 2011 WL 893506 at Findings of Fact ¶ 39 (Minn. Dist. Ct. 2011).

52 "Nadia's Ottawa 'Hell' Described."

53 Criminal Code of Canada, R.S. 1985, c-46, s.241, provides that "Every one who (*a*) counsels a person to commit suicide, or (*b*) aids or abets a person to commit suicide, whether suicide ensues or not, is guilty of an indictable offence and liable to imprisonment for a term not exceeding fourteen years."

54 Linda Nguyen, "MPs Back Motion on Suicide Bill; Legislation Clarifies Encouraging Suicide Online Is Illegal," *The Daily News* (Nanaimo, Canada), Nov. 19, 2009, at A12, www2.canada.com /nanaimodailynews/news/story.html?id=88ea0861-304d-49fe-8cd5-9564a9e16350.

55 Minn. Stat. Ann. §609.215 (West, Westlaw through the 2011 Regular Session).

56 *Minnesota v. Melchert-Dinkel*, 2011 WL 893506 (Minn. Dist. Ct. 2011).

57 *McCollum v. CBS, Inc.*, 202 Cal. App. 3d 989 (Cal. Ct. App. 1988). The lyrics were sung at one and a half times the normal rate of speech and in plaintiffs' words were not "immediately intelligible." Plaintiffs also argued that the music's pounding rhythm and hemisync sound waves impacted the listener's mental state.

58 Kim Murphy, "Suit over Suicide Is Tossed Out: No Proof That Rock Music Lyrics Led to It, Judge Says," *Los Angeles Times*, Aug. 8, 1986, at 3, http://articles.latimes.com/1986-08-08/news /mn-1834_1_suicide-solution.

59 *McCollum v. CBS, Inc.*, 202 Cal. App. 3d 989 (Cal. Ct. App. 1988).

60 Ibid. at 1001.

61 Ibid. The defendants claimed that the lyrics of "Suicide Solution" addressed the destructiveness of alcohol abuse. For example, the opening lines of the song state "Wine is fine but whiskey's quicker /Suicide is slow with liquor/Take a bottle drown your sorrows/Then it floods away tomorrows."

62 Ibid.

63 *DeFilippo v. National Broadcasting Co., Inc.*, 446 A.2d 1036 (R.I. 1982).

64 Ibid.

65 *Brandenburg v. Ohio*, 395 U.S. 444 (1969).

66 Ibid. at 447.

67 *Hess v. Indiana*, 414 U.S. 105, 108 (1973). Hess was convicted under Indiana's disorderly conduct statute for yelling "We'll take the fucking street later" during an anti-war demonstration.

68 Ibid.

69 Ibid. at 109.

70 *Minnesota v. Melchert-Dinkel*, 2011 WL 893506 at 33 (Minn. Dist. Ct. 2011).

71 Ibid. at 32.

72 *Snyder v. Phelps, Westboro Baptist Church*, 131 S. Ct. 1207 (2011).

73 Ibid. at 1217.

74 *Minnesota v. Melchert-Dinkel*, 2011 WL 893506 at 34 (Minn. Dist. Ct. 2011).

75 Minn. Stat. Ann. §609.215 (West, Westlaw through the 2011 Regular Session).

76 Warrant of Commitment, *Minnesota v. Melchert-Dinkel*, Case No. 66-CR-10-1193 (Minn. Dist. Ct. May 4, 2011).

77 Ibid. This adds 40 days to the 320-day prison sentence. Melchert-Dinkel will be on probation for 15 years and be prohibited from working with "vulnerable adults" or in a health care field.

78 Ibid. The order also imposes over $18,000 in fines and a $29,450 restitution award to Kajouji's family.

79 *Snyder v. Phelps, Westboro Baptist Church*, 131 S. Ct. 1207, 1229 n.1 (2011).

80 Ibid.

81 *People v. Dominguez*, 64 Cal. Rptr. 290 (Cal. Ct. App. 1967).

82 *People v. Pointer*, 199 Cal. Rptr. 357 (Cal. Ct. App. 1984) (a woman convicted of child endangerment and violation of a child custody decree was ordered not to conceive during the five-year probationary period; the condition of probation was overturned); *State v. Mosburg*, 768 P.2d 313 (Kan. Ct. App. 1989) (a woman convicted of endangering a child was paroled on the condition she not become pregnant; reversed); *State v. Livingston*, 372 N.E.2d 1335 (Ohio Ct. App. 1976) (striking the probation condition that a defendant convicted of child abuse not have a child for a period of five years).

83 Sameer Hinduja and Justin W. Patchin, "Cyberbullying Victimization," Cyberbullying Research Center, 2010, www.cyberbullying.us/2010_charts/cyberbullying_victim_2010.jpg.

84 Ibid.

85 "Police: Facebook Site May Have Led to Beating of 12-Year-Old," Nov. 22, 2009, www.cnn.com/2009/US/11/22/california.redhead.attack.facebook/index.html.

86 Bradley Schlegel, "Mother Warns Souderton Area Parents of Dangers of Cyberbullying," April 20, 2011, www.montgomerynews.com/articles/2011/04/20/souderton_independent/news/doc4dacfd60db530368715652.txt; Mike Celizic, "MySpace Victim's Mom Disappointed by Ruling," July 3, 2009, http://today.msnbc.msn.com/id/31722986/ns/today_people/.

87 Christopher Maag, "A Hoax Turned Fatal Draws Anger but No Charges," *The New York Times*, Nov. 28, 2007, at A23, www.nytimes.com/2007/11/28/us/28hoax.html?ref=meganmeier.

88 Brief of *Amici Curiae* Electronic Frontier Foundation in Support of Defendant's Motion to Dismiss Indictment for Failure to State an Offense and For Vagueness at 3, *U.S. v. Drew*, CR-08-0582-GW (C.D. Cal. Sep. 4, 2008).

89 *U.S. v. Drew*, 259 F.R.D. 449, 452 (C.D. Cal. 2009).

90 Maag, "A Hoax Turned Fatal Draws Anger but No Charges."

91 Celizic, "MySpace Victim's Mom Disappointed by Ruling."

92 Jennifer Steinhauer, "Verdict in MySpace Suicide Case," *The New York Times*, Nov. 27, 2008, at A25, www.nytimes.com/2008/11/27/us/27myspace.html.

93 *See*, Mo. Ann. Stat. §565.090.

94 Steinhauer, "Verdict in MySpace Suicide Case."

95 Kim Zitter, "Jurors Wanted to Convict Lori Drew of Felonies but Lacked Evidence," Dec. 1, 2008, www.wired.com/threatlevel/2008/12/jurors-wanted-t/.

96 Brian Kozlowski, "Lori Drew Convicted on Three Misdemeanor Counts of Violating MySpace Terms of Service in 'Cyberbullying' Case," JOLT Digest, Dec. 4, 2008, http://jolt.law.harvard.edu/digest/telecommunications/united-states-v-drew-2. JOLT Digest is an online companion to the *Harvard Journal of Law and Technology*.

97 Zitter, "Jurors Wanted to Convict Lori Drew of Felonies but Lacked Evidence."

98 Ibid.

99 *U.S. v. Drew*, 259 F.R.D. 449, 466 (C.D. Cal. 2009).

100 Vera Ranieri, "Conviction in Lori Drew MySpace Case Thrown Out," JOLT Digest, Sep. 4, 2009, http://jolt.law.harvard.edu/digest/9th-circuit/united-states-v-drew-3.

101 Celizic, "MySpace Victim's Mom Disappointed by Ruling."

102   M. F. Hertz and C. David-Ferdon, "Electronic Media and Youth Violence: A CDC Issue Brief for Educators and Caregivers," Centers for Disease Control and Prevention, 2008, www.cdc.gov /ViolencePrevention/pdf/EA-brief-a.pdf.

103   Mo. Ann. Stat. §565.090.

104   Lance Whitney, "Missouri Woman Charged with Cyberbullying," Aug. 19, 2009, http://news.cnet .com/8301-13578_3-10313304-38.html.

105   Steven Pokin, "St. Peters Woman Found Not Guilty in Cyber Harassment Case," *St. Charles Journal*, Feb. 17, 2011, www.stltoday.com/suburban-journals/stcharles/news/article_6dac3355-56dc-505e -922c-9d62b7e1080f.html.

106   Shane Anthony, "Jury Finds Woman Not Guilty in Cyber Harassment Case," Feb. 18, 2011, www .stltoday.com/news/local/stcharles/article_756d7f37-0088-5b47-b64f0cda15312fbb.html.

107   Pokin, "St. Peters Woman Found Not Guilty in Cyber Harassment Case."

108   Anthony, "Jury Finds Woman Not Guilty in Cyber Harassment Case."

109   Lance Whitney, "Missouri Woman Charged with Cyberbullying," Aug. 19, 2009, http://news.cnet .com/8301-13578_3-10313304-38.html.

110   "Mo. Woman Acquitted of Cyber Harassment," Feb. 18, 2011, www.nbcactionnews.com/dpp/news /state/missouri/mo.-woman-acquitted-of-cyber-harassment.

111   Pokin, "St. Peters Woman Found Not Guilty in Cyber Harassment Case."

112   Robert Meyer and Michel Cukier, "Assessing the Attack Threat Due to IRC Chat," in *Proceedings of the International Conference on Dependable Systems and Networks DSN06*, Philadelphia, Pa., June 25–28, 2006, www.enre.umd.edu/content/rmeyer-assessing.pdf.

113   "Working to Halt Online Abuse: 2010 Cyberstalking Statistics," www.haltabuse.org/resources /stats/2010Statistics.pdf. The group defines cyberharassment as "a course of conduct directed at a specific person that causes substantial emotional distress in such person and serves no legitimate purpose or words, gestures, and actions which tend to annoy, alarm and abuse verbally another person [and the offender has been asked to stop]."

114   Danielle Keats Citron, "Cyber Civil Rights," 89 *Boston University Law Review* 61, 64 (2009).

115   Danielle Keats Citron, "Civil Rights in Our Information Age," in *The Offensive Internet: Privacy, Speech, and Reputation*, ed. Saul Levmore and Martha C. Nussbaum (Cambridge, Mass.: Harvard University Press, 2010), 31–49, 36.

116   Cheryl Lindsey Seelhoff, "A Chilling Effect: The Oppression and Silencing of Women Journalists and Bloggers Worldwide," 37 *Off Our Backs* 18, 20 (2007).

117   Alex Pham, "Cyber-bullies' Abuse, Threats Hurl Fear into the Blogosphere," *Los Angeles Times*, March 31, 2007, at C1, http://articles.latimes.com/2007/mar/31/business/fi-internet31.

118   Creating Passionate Users blog, http://headrush.typepad.com/about.html.

119   Pham, "Cyber-bullies' Abuse, Threats Hurl Fear into the Blogosphere."

120   Dylan Tweney, "Kathy Sierra Case: Few Clues, Little Evidence, Much Controversy," April 16, 2007, www.wired.com/techbiz/people/news/2007/04/kathysierra; "Cathy Seipp and Kathy Sierra: Wild in the Blog-o-sphere," Fishbowl LA blog, March 27, 2007, www.mediabistro.com/fishbowlla/cathy -seipp-and-kathy-sierra-wild-in-the-blog-o-sphere_b3903 (screenshot of posted picture).

121   Tweney, "Kathy Sierra Case."

122   Pham, "Cyber-bullies' Abuse, Threats Hurl Fear into the Blogosphere."

123   "Blog Death Threats Spark Debate," March 27, 2007, http://news.bbc.co.uk/go/pr/fr/-/2/hi/ technology/6499095.stm.

124   Tweney, "Kathy Sierra Case"; Dan Fost, "Bad Behavior in the Blogosphere: Vitriolic Comments Aimed at Tech Writer Make Some Worry About Downside of Anonymity," *San Francisco Chronicle*,

March 29, 2007, at A1, http://articles.sfgate.com/2007-03-29/news/20870853_1_david-sifry-bloggers -free-speech.

125 Kathy Sierra and Chris Locke, "Coordinated Statements on the Recent Events," April 1, 2007, www .rageboy.com/statements-sierra-locke.html.

126 Creating Passionate Users blog, http://headrush.typepad.com.

127 "Kathy Sierra: Author, Blogger, Creating Passionate Users," http://en.oreilly.com/et2008/public /schedule/speaker/2227.

128 Heather Havenstein, "Q&A: Death Threats Force Blogger to Sidelines," March 27, 2007, www .computerworld.com/s/article/9014647/Q_A_Death_Threats_Force_Blogger_to_Sidelines?taxonom yId=16&pageNumber=2.

129 *See, e.g.,* "Kathy Sierra Author, Blogger, Creating Passionate Users," scheduled speech for May 26, 2010, www.gov2expo.com/gov2expo2010/public/schedule/speaker/2227.

130 First Amended Complaint at ¶¶ 7, 20, 21, *Doe I et al. v. Individuals,* No. 307CV00909 (D. Conn. Nov. 8, 2007).

131 Ibid. at ¶¶ 20, 21.

132 Ibid. at ¶ 21.

133 Ibid. at ¶ 21.

134 Ibid. at ¶ 20.

135 David Margolick, "Slimed Online: Two Lawyers Fight Cyberbullying," Feb. 11, 2009, www.portfolio .com/news-markets/national-news/portfolio/2009/02/11/Two-Lawyers-Fight-Cyber-Bullying/index .html.

136 First Amended Complaint at ¶ 15, *Doe I et al. v. Individuals,* No. 307CV00909 (D. Conn. Nov. 8, 2007) (citing 2007 statistics).

137 Ibid. at ¶ 12.

138 Margolick, "Slimed Online."

139 Brian Leiter, "Penn Law Student, Anthony Ciolli, Admits to Running Prelaw Discussion Board Awash in Racist, Anti-Semitic, Sexist Abuse," Leiter Reports: A Philosophy Blog, March 11, 2005, http://leiterreports.typepad.com/blog/2005/03/penn_law_studen.html.

140 Ibid.

141 First Amended Complaint at ¶¶ 21, 31, *Doe I et al. v. Individuals,* No. 307CV00909 (D. Conn. Nov. 8, 2007).

142 Ibid. at ¶ 21.

143 Margolick, "Slimed Online."

144 First Amended Complaint at ¶ 21, *Doe I et al. v. Individuals,* No. 307CV00909 (D. Conn. Nov. 8, 2007).

145 "Heide Motaghi Iravani Biography," Cleary Gottlieb, www.cgsh.com/hiravani/ (this is the website of the New York law firm at which she now works).

146 First Amended Complaint at ¶ 21, *Doe I et al. v. Individuals,* No. 307CV00909 (D. Conn. Nov. 8, 2007).

147 Ibid. at ¶¶ 45, 46.

148 Ibid. at ¶ 46.

149 Ibid. at ¶ 45.

150 Ibid. at ¶ 15.

151 Margolick, "Slimed Online."

152 First Amended Complaint at ¶ 61, *Doe I et al. v. Individuals*, No. 307CV00909 (D. Conn. Nov. 8, 2007).

153 Ibid. at ¶ 18.

154 Margolick, "Slimed Online."

155 First Amended Complaint at ¶ 22, *Doe I et al. v. Individuals*, No. 307CV00909 (D. Conn. Nov. 8, 2007).

156 Ibid. at ¶ 35.

157 Ibid. at ¶ 35.

158 Ibid. at ¶ 35; Margolick, "Slimed Online."

159 Margolick, "Slimed Online."

160 Complaint, *Doe I et al. v. Ciolli et al.*, No. 307CV00909 (D. Conn. June 8, 2007).

161 Order Denying Defendant's Motion to Quash Subpoena and Motion to Proceed Anonymously, *Doe I et al. v. Individuals*, No. 307CV00909 (D. Conn. June 13, 2008).

162 Ibid.

163 Ibid.

164 Margolick, "Slimed Online."

165 First Amended Complaint at ¶ 22, *Doe I et al. v. Individuals*, No. 307CV00909 (D. Conn. Nov. 8, 2007).

166 *See, e.g.*, Notice of Settlement and Request for Dismissal of Action Against Defendant "D," *Doe I et al. v. Individuals*, No. 307CV00909 (D. Conn. Sep. 9, 2009).

167 June 7, 2011, Thread, www.autoadmit.com/thread.php?thread_id=1667558&mc=4&forum_id=2#18185964; June 6, 2011, Thread, www.autoadmit.com/thread.php?thread_id=1667187&mc=3&forum_id=2.

168 "ALEP's New Postdoctoral Fellow—Brittan Heller," Afghanistan Legal Education Project, Aug. 15, 2010, http://alep.stanford.edu/?p=403.

169 Sep. 25, 2010, Thread, www.autoadmit.com/thread.php?thread_id=1433716&mc=3&forum_id=2; Oct. 6, 2010, Thread, www.autoadmit.com/thread.php?thread_id=1443743&mc=1&forum_id=2.

170 Citron, "Civil Rights in Our Information Age," 39.

171 "Death Online."

172 Kevin Poulsen, "Dangerous Japanese 'Detergent Suicide' Technique Creeps into U.S.," March 13, 2009, www.wired.com/threatlevel/2009/03/japanese-deterg/#.

173 "Church of Euthanasia FAQ," www.churchofeuthanasia.org/coefaq.html. The Church of Euthanasia states that it is "devoted to restoring balance between Humans and the remaining species on Earth" and that its motto is "Save the Planet, Kill Yourself," and it claims to support voluntary forms of population reduction.

174 Examples include *The Complete Manual of Suicide* by Frederic P. Miller, described on www.barnesandnoble.com as a 198-page book providing "explicit descriptions and analysis on a wide range of suicide methods," noting that most of the content is available for free online and that the book "is not a suicide manual for the terminally ill"; *The Peaceful Pill Handbook* by Philip Haig Nitschke, described on www.amazon.com as "an easy-to-understand overview of practical end-of-life options. . . . aimed at Seniors & folk who are seriously ill," including a unique Reliability and Peacefulness Test for methods comparison and photos of updated drug labeling; and *Suicide and Attempted Suicide: Methods and Consequences* by Geo Stone, described on www.amazon.com as "essentially a guide on how to commit suicide."

175 Complaint at ¶ 106, *Ciolli v. Iravani et al.*, No. 2:08-CV-02601 (E.D. Pa. March 4, 2008).

176   First Amended Complaint, *Doe I et al. v. Individuals*, No. 307CV00909 (D. Conn. Nov. 8, 2007).

177   Complaint at ¶ 24, *Ciolli v. Iravani et al.*, No. 2:08-CV-02601 (E.D. Pa. March 4, 2008).

178   Order Dismissing Action with Prejudice, *Ciolli v. Iravani et al.*, No. 2:08-CV-02601 (E.D. Pa. Nov. 23, 2009).

179   Amir Efrati, "Law Firm Rescinds Offer to Ex-AutoAdmit Executive," The Wall Street Journal law blog, May 3, 2007, http://blogs.wsj.com/law/2007/05/03/law-firm-rescinds-offer-to-ex-autoadmit -director/.

180   Ibid.

181   *Chicago Lawyers' Committee for Civil Rights Under Law, Inc. v. Craigslist, Inc.*, 519 F.3d 666 (7th Cir. 2008).

182   *Fair Housing Council of San Fernando Valley v. Roommates.com, LLC*, 521 F.3d 1157 (9th Cir. 2008).

183   Ibid.

184   C. Johnson, "Radio Station Found Liable in Water Intoxication Death Suit," Oct. 30, 2009, www .news10.net/news/story.aspx?storyid=69570.

185   Nancy S. Kim, "Web Site Proprietorship and Online Harassment," 2009 *Utah Law Review* 993, 998 (2009).

186   Ibid. at 1020.

187   Sam Bayard, "New Jersey Prosecutors Set Sights on JuicyCampus," Citizen Media Law Project, March 21, 2008, www.citmedialaw.org/blog/2008/new-jersey-prosecutors-set-sights-juicycampus.

188   Michael Arrington, "What Exactly Did the JuicyCampus Founder Think Would Happen?," March 2, 2008, http://techcrunch.com/2008/03/02/what-exactly-did-the-juicycampus-founder-think-would -happen/; Daniel Solove, "JuicyCampus: The Latest Breed of Gossip Website," Dec. 9, 2007, www .concurringopinions.com/archives/2007/12/juicy_campus_th.html.

8. Privacy of Place

1   William Bender, "Parents Meet to Slam Lower Merion Spy-Cam Suit," March 3, 2010, http://articles .philly.com/2010-03-03/news/24956833_1_class-action-status-class-action-parents.

2   L-3 Services, Inc., "Initial LANrev System Findings," *Lower Merion School District Forensics Analysis Part 1*, May 2010, at 7.

3   Ballard Spahr, "Report of Independent Investigation: Regarding Remote Monitoring of Student Laptop Computers by the Lower Merion School District," May 3, 2010, at 35.

4   Joseph Tanfani, "How a Lawsuit over School Laptops Evolved," *The Philadelphia Inquirer*, March 21, 2010, at A01, published online as "How School Web Cam Debacle Evolved," http://articles.philly .com/2010-03-21/news/25215619_1_web-cam-computer-files-school-board-member; Brian Sweeney, "The Top 50 School Districts 2008," *Philadelphia Magazine*, Aug. 28, 2008, www.phillymag.com /articles/the_top_50_school_districts_2008/; 2008 Ranking of 105 School Districts spreadsheet, *Philadelphia Magazine*, www.phillymag.com/files/images/Full_ranking.pdf; Ballard Spahr, "Report of Independent Investigation," at 1.

5   Complaint at 5-6, *Robbins v. Lower Merion School District*, No. 2:10-CV-00665 (E.D. Pa. Feb. 16, 2010).

6   Ballard Spahr, "Report of Independent Investigation," at 1.

7   John P. Martin, "Lower Merion School Board to Consider Webcam Policy," *The Philadelphia Inquirer*, July 19, 2010, at A01, http://articles.philly.com/2010-07-19/news/24970664_1_webcam -laptops-students; Chloe Albanesius, "Another Lawsuit Filed over School Webcam Spying," July 30, 2010, www.pcmag.com/article2/0,2817,2367209,00.asp; Larry Magid, "Students'-Eye View of Webcam Spy Case," Feb. 19, 2010, http://news.cnet.com/8301-30977_3-10457077-10347072.html;

Complaint at 5, *Robbins v. Lower Merion School District*, No. 2:10-CV-00665 (E.D. Pa. Feb. 16, 2010).

8   Ibid.

9   Michael Smerconish, "Web Cam Violated Third-Party Rights," April 25, 2010, www.smerconish .com/pages/pages.php?page=98.

10 John P. Martin, "1,000s of Web Cam Images, Suit Says," *The Philadelphia Inquirer*, April 16, 2010, at A01, published online on April 15, 2010, as "Lawyer: Laptops Took Thousands of Images," www .philly.com/philly/news/year-in-review/20100415_Lawyer__Laptops_took_thousands_of_photos .html.

11 Ballard Spahr, "Report of Independent Investigation," at 1.

12 Smerconish, "Web Cam Violated Third-Party Rights."

13 Magid, "Students'-Eye View of Webcam Spy Case."

14 Ballard Spahr, "Report of Independent Investigation," at 91.

15 Ibid. at 2.

16 John P. Martin, "U.S. Ends Webcam Probe; No Charges," Aug. 17, 2010, http://articles.philly .com/2010-08-17/news/24973417_1_laptops-webcam-reasonable-doubt.

17 Tony Romm, "Specter Introduces Anti-Video Surveillance Bill Following Pa. Laptop Camera Flap," April 16, 2010, http://thehill.com/blogs/hillicon-valley/technology/92667-specter-introduces -anti-video-surveillance-bill-following-pa-laptop-camera-flap.

18 S. 3214, 111th Cong. (2010).

19 Romm, "Specter Introduces Anti-Video Surveillance Bill Following Pa. Laptop Camera Flap."

20 Complaint, *Robbins v. Lower Merion School District*, No. 2:10-CV-00665 (E.D. Pa. Feb. 11, 2010).

21 Ibid. at ¶¶ 27–39.

22 Ibid. at ¶¶ 49–55.

23 L-3 Services, Inc., "Initial LANrev System Findings," at 16.

24 Derrick Nunnally, "Second Suit over Lower Merion Webcam Snooping," *The Philadelphia Inquirer*, July 28, 2010, at A01, http://articles.philly.com/2010-07-28/news/24972175_1_webcam-monitoring -system-culinary-school; Complaint at 7, *Hasan v. Lower Merion School District*, No. 2:10-CV-03663 (E.D. Pa. July 27, 2010).

25 Nunnally, "Second Suit over Lower Merion Webcam Snooping."

26 Ibid.

27 Complaint at 7, *Hasan v. Lower Merion School District*, No. 2:10-CV-03663 (E.D. Pa. July 27, 2010).

28 Nunnally, "Second Suit over Lower Merion Webcam Snooping."

29 William Bender, "Parents Meet to Slam Lower Merion Spy-Cam Suit," *The Philadelphia Daily News*, March 3, 2010, at 6, http://articles.philly.com/2010-03-03/news/24956833_1_class-action -status-class-action-parents.

30 Brief of Amicus Curiae American Civil Liberties Union of Pennsylvania Support Issuance of Injunction at 2, *Robbins v. Lower Merion School District*, No. 2:10-CV-00665 (E.D. Pa. Feb. 22, 2010).

31 Ibid., *citing U.S. v. Zimmerman*, 277 F.3d 426, 431 (3d Cir. 2002).

32 Ibid., *citing Groh v. Ramirez*, 540 U.S. 551, 559 (2004) (citations omitted).

33 Ballard Spahr, "Report of Independent Investigation," at 5, 57–58.

34 Dan Hardy, Derrick Nunnally, and John Shiffman, "Laptop Camera Snapped Away in One Classroom," *Philadelphia Inquirer*, Feb. 22, 2010, at A01, http://articles.philly.com/2010-02-22 /news/25218854_1_school-issued-laptop-photos-apple-macbook.

35   Ibid.

36   Ballard Spahr, "Report of Independent Investigation," at 29, *citing* Email from V. DiMedio to Student Intern, Aug. 11, 2008, App. Tab 41.

37   Ballard Spahr, "Report of Independent Investigation," at 28, *citing* Email from Student Intern to M. Perbix, Aug. 11, 2008, App. Tab 42.

38   Ibid. at 42.

39   Ibid. at 62.

40   Order Granting Injunctive Relief, *Robbins v. Lower Merion School District*, No. 2:10-CV-00665 (E.D. Pa. May 14, 2010).

41   "Lower Merion School District Settles Webcam Spying Lawsuits For $610,000," Oct. 11, 2010, www.huffingtonpost.com/2010/10/11/lower-merion-school-distr_n_758882.html; Order Granting Plaintiff's Petition for Leave to Settle, *Robbins v. Lower Merion School District*, No. 2:10-CV-03663-JD (E.D. Pa. Oct. 14, 2010). The settlement called for Robbins to get $175,000, for Hasan to get $10,000, and for their lawyer, Mark Haltzman, to get $425,000 for his work on the case.

42   Ed Pilkington, "Tyler Clementi, Student Outed as Gay on Internet, Jumps to His Death," Sep. 30, 2010, www.guardian.co.uk/world/2010/sep/30/tyler-clementi-gay-student-suicide.

43   David Lohr, "Did Tyler Clementi Reach Out for Help Before Suicide?," Sep. 30, 2010, www.aolnews.com/2010/09/30/did-tyler-clementi-reach-out-for-help-before-suicide/.

44   Emily Friedman, "Victim of Secret Dorm Sex Tape Posts Facebook Goodbye, Jumps to His Death," Sep. 29, 2010, http://abcnews.go.com/US/victim-secret-dorm-sex-tape-commits-suicide/story?id=11758716&page=1.

45   Paul Thompson, "'He Was Spying on Me': Student Who Killed Himself after Secret Gay Sex Film Made Desperate Cry for Help on Day of His Death," Oct. 1, 2010, www.dailymail.co.uk/news/article-1316600/Tyler-Clementi-suicide-Emails-reveal-teenagers-private-torment.html.

46   Friedman, "Victim of Secret Dorm Sex Tape Posts Facebook Goodbye, Jumps to His Death."

47   Pilkington, "Tyler Clementi, Student Outed as Gay on Internet, Jumps to His Death."

48   Friedman, "Victim of Secret Dorm Sex Tape Posts Facebook Goodbye, Jumps to His Death."

49   Nicole Weisensee Egan, "Student Gets Leniency in Rutgers Webcam-Spying Case," May 6, 2011, www.people.com/people/article/0,,20487550,00.html.

50   Kayla Webley, "Former Rutgers Student Dharun Ravi Indicted in Tyler Clementi Suicide," April 20, 2011, http://newsfeed.time.com/2011/04/20/former-rutgers-student-dharun-ravi-indicted-in-tyler-clementi-suicide/.

51   Richard Perez-Pena and Nate Schweber, "Roommate Is Arraigned in Rutgers Suicide Case," *The New York Times*, May 23, 2011, at A22, www.nytimes.com/2011/05/24/nyregion/roommate-arraigned-in-rutgers-spy-suicide-case.html.

52   John Schwartz, "As Big PC Brother Watches, Users Encounter Frustration," *The New York Times*, Sep. 5, 2001, at C6, www.nytimes.com/2001/09/05/business/as-big-pc-brother-watches-users-encounter-frustration.html?src=pm.

53   "A British Spy, a Speedo and Facebook," Public Radio International, July 8, 2009, www.pri.org/science/technology/british-spy-facebook1476.html.

54   "Spy Job: Britain's New MI6 Chief Revealed," June 6, 2009, http://news.sky.com/skynews/Home/UK-News/Britains-UN-Ambassador-Sir-John-Sawers-Is-To-Be-New-Head-Of-MI6-Replacing-Sir-John-Scarlett/Article/200906315309448.

55   Kim LaCapria, "Facebook Places Maybe Nets Robbers $100K in Booty in New Hampshire," Sep. 10, 2010, www.inquisitr.com/84467/facebook-places-robbery/; Casey Chan, "Robbers Checked Facebook Status Updates to See When People Weren't Home," Sep. 12, 2010, http://gizmodo.com/5636025/robbers-used-facebook-to-see-when-people-werent-home; Nick Bilton, "Burglars Said

to Have Picked Houses Based on Facebook Updates," The New York Times Bits blog, Sep. 12, 2010, http://bits.blogs.nytimes.com/2010/09/12/burglars-picked-houses-based-on-facebook-updates/.

56   Kate Murphy, "Web Photos That Reveal Secrets, like Where You Live," *The New York Times*, Aug. 12, 2010, at B6, www.nytimes.com/2010/08/12/technology/personaltech/12basics.html.

57   "Facebook to Release Phone Numbers, Addresses to Third-Party Developers," http://truthmovement .com/?p=2070; Bianca Bosker, "Facebook to Share Users' Home Addresses, Phone Numbers with External Sites," Feb. 28, 2011, www.huffingtonpost.com/2011/02/28/facebook-home-addresses -phone-numbers_n_829459.html.

58   Letter from U.S. Representatives Edward Markey and Joe Barton to Mark Zuckerberg, Feb. 2, 2011.

59   Letter from Marne Levine, Vice President of Global Public Policy, Facebook, to Representatives Markey and Barton, Feb. 23, 2011.

60   Shane Anthony, "Jury Finds Woman Not Guilty in Cyber Harassment Case," Feb. 18, 2011, www .stltoday.com/news/local/stcharles/article_756d7f37-0088-5b47-b64f-0cda15312fbb.html.

9. Privacy of Information

1   "Did the Internet Kill Privacy?," Feb. 6, 2011, www.cbsnews.com/stories/2011/02/06/sunday /main7323148.shtml.

2   Jaime Sarrio, "Ex-Teacher Fighting to Get Back into Classroom," *The Atlanta Journal-Constitution*, Nov. 10, 2010, at B4.

3   Carman Peterson, "Principal Receives Threats After News Coverage of Former Teacher's Lawsuit," Nov. 18, 2009, http://beta.barrowcountynews.com/archives/5020/.

4   Maureen Downey, "Barrow Teacher Done in By Anonymous 'Parent' E-mail About Her Facebook Page," Atlanta Journal-Constitution blog, Nov. 13, 2009, http://blogs.ajc.com/get-schooled-blog/2009 /11/13/barrow-teacher-done-in-by-anonymous-email-with-perfect-punctuation/.

5   Ibid.

6   Kristi Reed, "BCS Responds to Teacher Lawsuit," Nov. 12, 2009, www.barrowjournal.com/archives /1970-BCS-responds-to-teacher-lawsuit.html#extended.

7   Kristi Reed, "Hearing Set in Facebook Case," April 24, 2010, www.barrowjournal.com/archives/2775 -Hearing-set-in-Facebook-case.html.

8   Kaitlin Madden, "12 Ways to Get Fired for Facebook," Aug. 9, 2010, www.careerbuilder.ca/Article /CB-606-Workplace-Issues-12-Ways-to-Get-Fired-for-Facebook/.

9   Sarrio, "Ex-Teacher Fighting to Get Back into Classroom." The decision occurred in fall 2010.

10   Downey, "Barrow Teacher Done In by Anonymous 'Parent' E-mail About Her Facebook Page."

11   Ibid.

12   Ibid.

13   Kaplan Press Release, "At Top Schools, One in Ten College Admissions Officers Visits Applicants' Social Networking Sites," Sep. 18, 2008, www.kaplan.com/SiteCollectionDocuments/Kaplan.com /PressReleases/2007/Sep.%2018%20-%20CAO%20survey%20results.pdf.

14   Ibid.

15   Amanda Ricker, "City Requires Facebook Passwords from Job Applicants," June 18, 2009, www .bozemandailychronicle.com/news/article_a9458e22-498a-5b71-b07d-6628b487f797.html.

16   Matt Gouras, "City Drops Request for Internet Passwords," June 19, 2009, www.msnbc.msn.com /id/31446037/ns/technology_and_science-security/t/city-drops-request-internet-passwords/.

17   Letter from Deborah A. Jeon, Legal Director, ACLU Maryland, to Secretary Gary D. Maynard, Maryland Department of Public Safety and Correctional Services, Jan. 25, 2011, www.aclu-md.org /aPress/Press2011/collinsletterfinal.pdf.

18   Nick Madigan, "Officer Says He Had to Give Facebook Password for Job," *The Baltimore Sun*, at 3A, published online on Feb. 23, 2011, as "Officer Forced to Reveal Facebook Page," http://articles .baltimoresun.com/2011-02-23/news/bs-md-ci-officer-facebook-password-20110223_1_facebook -page-facebook-password-privacy-protections.

19   Meta Pettus, "Man Refuses to Give Facebook Password for Background Check," Feb. 23, 2011, www .wusa9.com/news/local/story.aspx?storyid=137495&catid=158.

20   Madigan, "Officer Says He Had to Give Facebook Password for Job."

21   Monica Hayes, "Md. Corrections No Longer Requiring Social Media Passwords," April 6, 2011, www.nbcwashington.com/news/local/MARYLAND-AGENCY-CHANGES-POLICY-ON -FACEBOOK-119361764.html.

22   "ACLU Says Division of Corrections Revised Social Media Policy Remains Coercive and Violates 'Friends' Privacy Rights," April 18, 2011, www.aclu.org/technology-and-liberty/aclu-says-division -corrections-revised-social-media-policy-remains-coercive-a.

23   Nadia Wynter, "'Gaydar' Project at MIT Attempts to Predict Sexuality Based on Facebook Profiles," Sep. 22, 2009, www.nydailynews.com/lifestyle/2009/09/22/2009-09-22_gaydar_project_at_mit _attempts_to_predict_sexuality_based_on_facebook_profiles.html; Carter Jernigan and Behram F. T. Mistree, "Gaydar: Facebook Friendships Expose Sexual Orientation," 14 *First Monday* (online) (2009).

24   Kashmir Hill, "Fitbit Moves Quickly After Users' Sex Stats Exposed," July 5, 2011, http://blogs.forbes .com/kashmirhill/2011/07/05/fitbit-moves-quickly-after-users-sex-stats-exposed/.

25   Ibid.

26   Leena Rao, "Sexual Activity Tracked by Fitbit Shows Up in Google Search Results," July 3, 2011, http://techcrunch.com/2011/07/03/sexual-activity-tracked-by-fitbit-shows-up-in-google-search -results/.

27   Ki Mae Heussner, "Teacher Loses Job After Commenting About Students, Parents on Facebook," Aug. 19, 2010, http://abcnews.go.com/Technology/facebook-firing-teacher-loses-job-commenting -students-parents/story?id=11437248.

28   *Ledbetter v. Wal-Mart Stores, Inc.*, 2009 WL 1067018 (D. Colo. 2009).

29   *Romano v. Steelcase Inc.*, 907 N.Y.S.2d 650 (N.Y. Sup. Ct. 2010); *see also*, "Facebook as Evidence?," Sep. 27, 2010, www.nbcnewyork.com/news/local/Facebook-as-Evidence-103898924.html.

30   Order Denying in Part Motion to Reconsider Discovery Request, *Beye v. Horizon Blue Cross Blue Shield of New Jersey*, No. 06-05377 (D.N.J. Dec. 14, 2007).

31   *Hexum v. Hexum*, 719 N.W.2d 799, 4 (Wis. Ct. App. 2006).

32   *Carswell v. Carswell*, 2009 WL 3284874 at 5-6 (Conn. Super. Ct. 2009).

33   *Mackelprang v. Fidelity National Title Agency of Nevada, Inc.*, 2007 WL 119149 (D. Nev. 2007).

34   Joanne Kuzma, "Empirical Study of Privacy Issues Among Social Networking Sites," 6 *Journal of International Commercial Law and Technology*, 74, 76, 82 (2011).

35   Joseph Bonneau and Sören Preibusch, "The Privacy Jungle: On the Market for Data Protections in Social Networks," Eighth Workshop on the Economics of Information Security, London, United Kingdom, June 24–25, 2009; Joseph Bonneau and Sören Preibusch, "The Privacy Jungle: On the Market for Data Protections in Social Networks," in *Economics of Information Security and Privacy*, ed. Tyler Moore, David J. Pym, and Christos Ioannidis (New York: Springer, 2010), 121– 167.

36   Ibid.

37   Eliot Van Buskirk, "Report: Facebook CEO Mark Zuckerberg Doesn't Believe in Privacy," April 28, 2010, www.wired.com/epicenter/2010/04/report-facebook-ceo-mark-zuckerberg-doesnt-believe -in-privacy.

38   Ryan Singel, "Facebook's Gone Rogue; It's Time for an Open Alternative," May 7, 2010, www.wired
.com/epicenter/2010/05/facebook-rogue/.

39   Ibid.; In August 2011, Facebook announced that it would offer new privacy control options, including
a tool for users to limit which of their friends can view individual posts on Facebook, *see* Geoffrey A.
Fowler, "What Facebook's New Privacy Settings Mean for You," The Wall Street Journal Digits blog,
Aug.   23,   2011,   http://blogs.wsj.com/digits/2011/08/23/what-facebook%E2%80%99s-new-privacy
-settings-mean-for-you/.

40   "Facebook Privacy: A Bewildering Tangle of Options," May 12, 2010, www.nytimes.com/
interactive/2010/05/12/business/facebook-privacy.html.

41   Ibid.

42   Yen-Pei Huang, Dr. Tiong Goh, and Dr. Chern Li Lew, "Hunting Suicide Notes in Web 2.0—
Preliminary Findings," Ninth IEEE Symposium on Multimedia, 2007 Workshop, http://ieeexplore
.iece.org/stamp/stamp.jsp?arnumber=04476021.

43   Ibid.

44   *Olmstead v. United States*, 277 U.S. 438, 478 (1928) (Brandeis, J., dissenting).

45   Bobbie Johnson, "Privacy No Longer a Social Norm, says Facebook Founder," Jan. 11, 2010, www
.guardian.co.uk/technology/2010/jan/11/facebook-privacy.

46   *M.G. v. Time Warner, Inc.*, 89 Cal. App. 4th 623, 632 (Cal. Ct. App. 2001).

47   National Conference of State Legislatures, "Off-Duty Conduct," May 30, 2008, www.ncsl.org
/default.aspx?tabid=13369.

48   Minn. Stat. Ann. § 181.938; *see also* 820 Ill. Comp. Stat. Ann. 55/5; Mont. Code Ann. §§ 39-2-313,
39-2-314; Minn. Stat. Ann. § 181.938; N.C. Gen. Stat. § 95-28.2; Nev. Rev. Stat. Ann. § 613.333.

49   Robert Sprague, "Rethinking Information Privacy in an Age of Online Transparency," 25 *Hofstra
Labor and Employment Law Journal* 395 (2009).

50   Cal. Lab. Code §§ 96(k), 98.6; Colo. Rev. Stat. § 24-34-402.5; N.Y. Lab. Code § 201-d; N.D. Cent.
Code § 14-02/4-03,

51   Ibid.

52   N.Y. Lab. Code § 201-d.

53   *See City of Ontario v. Quon*, 130 S. Ct. 2619 (2010).

54   29 CFR §1635.8 (2010).

55   Mary Madden and Aaron Smith, "Reputation Management and Social Media," Pew Research Center,
May 26, 2010, http://pewinternet.org/~/media//Files/Reports/2010/PIP_Reputation_Management
_with_topline.pdf.

56   Charles Fried, "Privacy," 77 *Yale Law Journal* 475, 477 (1968).

57   Robert C. Post, "The Social Foundations of Privacy: Community and Self in the Common Law Tort,"
77 *California Law Review* 957, 1008–1009 (1989).

58   Jessica Bennett, "A Tragedy That Won't Fade Away," April 25, 2009, www.newsweek.com/2009/04/24
/a-tragedy-that-won-t-fade-away.html; Greg Hardesty, "High Court Declines to Review CHP
Catsouras Photo Case," April 15, 2010, http://articles.ocregister.com/2010-04-15/crime/24648927_1
_high-court-declines-nikki-catsouras-aaron-reich.

59   *Catsouras v. Department of California Highway Patrol*, 181 Cal. App. 4th 856 (Cal. Ct. App. 2010).

60   Jim Avila, Teri Whitcraft, and Scott Michels, "A Family's Nightmare: Accident Photos of Their Beautiful
Daughter Released," Nov. 16, 2007, http://abcnews.go.com/TheLaw/story?id=3872556&page=1.

61   *Catsouras v. Department of California Highway Patrol*, 181 Cal. App. 4th 856, 865, 866 (Cal. Ct. App.
2010).

62   Ibid.

63   Bennett, "A Tragedy That Won't Fade Away." Reich's attorney, Jon Schlueter, described the emailed images as a "a cautionary tale" and noted that "Any young person that sees these photos and is goaded into driving more cautiously or less recklessly—that's a public service."

64   Greg Hardesty, "Judge Dismisses Suit over CHP Photo Leak," March 21, 2008, http://articles .ocregister.com/2008-03-21/cities/24742374_1_chp-aaron-reich-dispatchers/2.

65   *Catsouras v. Department of California Highway Patrol*, 181 Cal. App. 4th 856, 863 (Cal. Ct. App. 2010).

66   Ibid. at 865.

67   Jessica Bennett, "At Long Last, a Small Justice," Feb. 5, 2010, www.newsweek.com/2010/02/04/at -long-last-a-small-justice.html.

68   Victoria Murphy Barret, "Anonymity & the Net," Oct. 15, 2007, www.forbes.com/forbes/2007 /1015/074.html.

69   Bennett, "At Long Last, a Small Justice."

70   Bennett, "A Tragedy That Won't Fade Away."

71   Barret, "Anonymity & the Net."

72   Avila et al., "A Family's Nightmare."

73   Barret, "Anonymity & the Net."

74   Ibid.

75   Avila et al., "A Family's Nightmare."

76   Ibid.

77   *Catsouras v. Department of California Highway Patrol*, 181 Cal. App. 4th 856 (Cal. Ct. App. 2010).

78   Ibid.; *see also*, Avila et al., "A Family's Nightmare."

79   *Catsouras v. Department of California Highway Patrol*, 181 Cal. App. 4th 856, 863, 864, 866 (Cal. Ct. App. 2010).

80   Bennett, "At Long Last, a Small Justice."

81   *Catsouras v. Department of California Highway Patrol*, 181 Cal. App. 4th 856, 864 (Cal. Ct. App. 2010).

82   Ibid. at 875.

10. FYI or TMI?: Social Networks and the Right to a Relationship with Your Children

1   *In re N.L.D.*, 344 S.W.3d 33, 38 (Tex. Ct. App. 2011).

2   "Big Surge in Social Networking Evidence Says Survey of Nation's Top Divorce Lawyers," American Academy of Matrimonial Lawyers, Feb. 10, 2010, www.aaml.org/about-the-academy/press/press -releases/e-discovery/big-surge-social-networking-evidence-says-survey-.

3   Nadine Brozan, "Divorce Lawyers' New Friend: Social Networks," *The New York Times,* May 13, 2011, at ST17, www.nytimes.com/2011/05/15/fashion/weddings /divorce-lawyers-new-friend-social -networks.html.

4   Ibid.

5   *Olson v. Olson*, 2010 WL 4517444 at 2 (Conn. Super. Ct. 2010).

6   Meghan Barr, "On Facebook, Wife Learns of Husband's 2nd Wedding," Aug. 5, 2010, www .huffingtonpost.com/2010/08/05/on-facebook-wife-learns-o_n_671556.html.

7   Sarina Fazan, "Facebook Unravels an Alleged Double Life of Prominent Tampa Business Man," Aug. 6, 2010, www.abcactionnews.com/dpp/news/region_tampa/facebook-unravels-an-alleged-double-life -of-prominent-tampa-business-man.

8   Edecio Martinez, "Lynn France Learns of Husband John France's 2nd Wedding . . . on Facebook," Aug. 5, 2010, www.cbsnews.com/8301-504083_162-20012786-504083.html.

9   Liz Brody, "Divorce, Courtesy Facebook. 5 Things You've Got to Know," July 25, 2010, http://shine .yahoo.com/channel/sex/divorce-courtesy-facebook-5-things-youve-got-to-know-2132630.

10 Erica Naone, "How Divorce Lawyers Use Social Networks," July 1, 2011, www.technologyreview .com/web/37943/?mod=chfeatured&a=f.

11 Caroline Black, "Baby with Bong Facebook Photo Arrested; Fla. Mom Rachel Stieringer Faces Drug Charges," Aug. 17, 2010, www.cbsnews.com/8301-504083_162-20013878-504083.html.

12 *Skinner v. Oklahoma*, 316 U.S. 535, 541 (1942).

13 *Stanley v. Illinois*, 405 U.S. 645, 651 (1976).

14 *Wisconsin v. Yoder*, 406 U.S. 205 (1972).

15 *Prince v. Massachusetts*, 321 U.S. 158, 166 (1944).

16 *Meyer v. Nebraska*, 262 U.S. 390, 402 (1923).

17 *Byrne v. Byrne*, 168 Misc. 2d 321, 322 (1996).

18 *In re T.T.*, 228 S.W.3d 312, 322-23 (Tex. Ct. App. 2007).

19 Steven Seidenberg, "Seduced: For Lawyers, the Appeal of Social Media Is Obvious. It's Also Dangerous," *American Bar Association Journal*, Feb. 1, 2011, www.abajournal.com/magazine/article/seduced_for _lawyers_the_appeal_of_social_media_is_obvious_dangerous/, quoting an instance mentioned by Linda Lea Viken, a family law specialist who heads the Viken Law Firm in Rapid City, S.D.

20 *Cedric D. v. Staciai W.*, 2007 WL 5515319 at 4 (Ariz. Ct. App. 2007).

21 Seidenberg, "Seduced."

22 Vesna Jaksic, "Litigation Clues Are Found on Facebook," *National Law Journal*, Oct. 15, 2007, at 1.

23 Sharon D. Nelson and John W. Simek, "Adultery in the Electronic Era: Spyware, Avatars, and Cybersex," 11 *Journal of Internet Law* 1, 2 (2008).

24 *Brown v. Crum*, 30 So. 3d 1254 (Miss. Ct. App. 2010).

25 *Smith v. Smith*, 2009 WL 195939 (Ark. Ct. App. 2009).

26 *Lipps v. Lipps*, 2010 WL 1379803 at 1 (Ark. Ct. App. 2010).

27 Ibid. at 1.

28 *High v. High*, 697 S.E.2d 690 (S.C. Ct. App. 2010).

29 *Pazdera v. Pazdera*, 794 N.W.2d 926 (Wis. Ct. App. 2010).

30 *Gainey v. Edington*, 24 So. 3d 333, 337 (Miss. Ct. App. 2009).

31 *In re Paternity of P.R.*, 940 N.E.2d 346, 347 (Ind. Ct. App. 2010).

32 *O'Brien v. O'Brien*, 899 So. 2d 1133, 1134 (Fla. Dist. Ct. App. 2005).

33 *Gurevich v. Gurevich*, 24 Misc. 3d 808 (N.Y. Sup. Ct. 2009).

34 *Miller v. Meyers*, 766 F. Supp. 2d 919 (W.D. Ark. 2011).

35 "Charges Dropped in Facebook Spy vs. Spy Case," June 9, 2011, www.thesmokinggun.com /documents/funny/facebook-spy-vs-spy-case-126493.

36 Ibid.

37 Alyson Shontell, "Woman Poses as a Sexy Teen on Facebook, Tricks Her Husband and Learns He's Trying to Kill Her," June 9, 2011, www.businessinsider.com/angel-david-voelkert-divorce-facebook -jessica-studebaker-murder-2011-6.

38 Ibid.

39 "Charges Dropped in Facebook Spy vs. Spy Case."

11. Social Networks and the Judicial System

1   Leila Atassi, "Cuyahoga County Judge Shirley Strickland Saffold Files $50 Million Lawsuit Against The Plain Dealer and Others," April 8, 2010, http://blog.cleveland.com/metro/2010/04/cuyahoga _county_judge_shirley.html.

2   Marci Stone, "Police: Two More Bodies Found at Anthony Sowell's Home in Cleveland," Nov. 3, 2009, www.examiner.com/headlines-in-salt-lake-city/police-two-more-bodies-found-at-anthony -sowell-s-home-cleveland.

3   Pierre Thomas, "10 Bodies in Sex Offender's Home: System Is Broken," Nov. 4, 2009, http://abcnews .go.com/WN/anthony-sowell-murder-case-highlights-broken-sex-offender/story?id=8999276; Soraya Roberts, "Anthony Sowell Rape and Murder Case Goes International," Nov. 9, 2009, http://articles .nydailynews.com/2009-11-09/news/17940661_1_camp-pendleton-anthony-sowell-bodies; Marci Stone, "Police: 10 Bodies Have Been Found in Anthony Sowell's Home and a Skull Bucket," Nov. 3, 2009, www.examiner.com/headlines-in-salt-lake-city/police-10-bodies-have-been-found-anthony -sowell-s-home-and-a-skull-bucket.

4   Helen Kennedy, "Suspected Ohio Serial Killer Anthony Sowell's Victim, Tanja Doss, Described Her Night of Terror," Nov. 5, 2009, http://articles.nydailynews.com/2009-11-05/news/17939011_1 _anthony-sowell-cops-bodies.

5   "Police Question Sowell in December After Rape Claim," Nov. 14, 2009, www.toledoblade.com /State/2009/11/14/Police-questioned-Sowell-in-December-after-rape-claim.html.

6   Thomas, "10 Bodies in Sex Offender's Home: System Is Broken."

7   Ibid.

8   "Pedestrian Hit by RTA Bus Dies," March 27, 2009, www.wkyc.com/news/local/story.aspx ?storyid=110049.

9   Brennan McCord and Eamon McNiff, "Judge Saffold Files $50M Suit Against Cleveland Newspaper over Online Comments," April 7, 2010, http://abcnews.go.com/TheLaw/cleveland-judge-denies -making-online-comments/story?id=10304420&page=1 (quotations omitted).

10   Atassi, "Cuyahoga County Judge Shirley Strickland Saffold Files $50 Million Lawsuit Against The Plain Dealer and Others"; James F. McCarty, "Lawyer for Anthony Sowell to Ask Judge Shirley Strickland Saffold to Step Down over Internet Comments," March 26, 2010, http://blog.cleveland .com/metro/2010/03/lawyer_for_anthony_sowell_to_a.html.

11   McCord and McNiff, "Judge Saffold Files $50M Suit Against Cleveland Newspaper over Online Comments."

12   McCarty, "Lawyer for Anthony Sowell to Ask Judge Shirley Strickland Saffold to Step Down over Internet Comments."

13   McCord and McNiff, "Judge Saffold Files $50M Suit Against Cleveland Newspaper over Online Comments."

14   "Heidi Klum Will Take Seal's Name," Oct. 6, 2009, www.popeater.com/2009/10/06/heidi-klum -name-change/.

15   James F. McCarty, "Anonymous Online Comments are Linked to the Personal Email Account of Cuyahoga County Common Pleas Judge Shirley Strickland Saffold," March 26, 2010, http://blog .cleveland.com/metro/2010/03/post_258.html.

16   Ibid.

17   McCord and McNiff, "Judge Saffold Files $50M Suit Against Cleveland Newspaper over Online Comments."

18   McCarty, "Anonymous Online Comments are Linked to the Personal Email Account of Cuyahoga County Common Pleas Judge Shirley Strickland Saffold."

19   Ibid.

20 "Saffolds Dismiss Lawsuit Against Plain Dealer, Settle with Advance Internet," Dec. 31, 2010, http:// blog.cleveland.com/metro/2010/12/saffolds_dismiss_lawsuit_again.html.

21 Atassi, "Cuyahoga County Judge Shirley Strickland Saffold Files $50 Million Lawsuit Against The Plain Dealer and Others."

22 Sowell was found guilty on July 22, 2011. "Anthony Sowell, 'The Cleveland Strangler,' Found Guilty of Murder," July 22, 2011, www.huffingtonpost.com/2011/07/21/anthony-sowell-guilty_n_905786 .html#s314099; Leila Atassi and Stan Donaldson, "GUILTY Reaction. Sowell Verdict a Relief to Victims' Families," Cleveland Plain Dealer, July 23, 2011, at A1, http://webmedia.newseum.org /newseum-multimedia/tpt/2011-07-23/pdf/OH_CPD.pdf.

23 Karen Farkas, "Judge Shirley Strickland Saffold Is Removed from the Anthony Sowell Murder Trial," April 22, 2010, http://blog.cleveland.com/metro/2010/04/judge_shirley_strickland_saffo_2.html.

24 Ibid.

25 Katheryn Hayes Tucker, "Judge Steps Down Following Questions About Facebook Relationship with Defendant," Florida Business Review (online), Jan. 8, 2010; Gena Slaughter and John G. Browning, "Social Networking Dos and Don'ts for Lawyers and Judges," 73 Texas Bar Journal 192, 194 (2010).

26 Stephani Gurr, "Habersham Judge Resigns Following Misconduct Allegations," Dec. 31, 2009, www.gainesvilletimes.com/archives/27787; resignation and photo of him with laptop in Blake Spurney, "Chief Judge Bucky Woods Resigns," Dec. 31, 2009, www.thenortheastgeorgian.com /articles/2010/01/03/news/top_stories/02topstory.txt.

27 Tucker, "Judge Steps Down."

28 Ibid.

29 Ibid.

30 Ibid.

31 Conference of Court Public Information Officers (CCPIO), "New Media and the Courts: The Current Status and a Look at the Future," Aug. 26, 2010, www.ccpio.org/documents/newmediaproject/New -Media-and-the-Courts-Report.pdf, 8.

32 American Bar Association, ABA's 2010 Legal Technology Survey Report.

33 Debra Cassens Weiss, "Blogging Assistant PD Accused of Revealing Secrets of Little-Disguised Clients," Sep. 10, 2009, www.abajournal.com/news/article/blogging_assistant_pd_accused_of _revealing_secrets_of_little-disguised_clie/.

34 In re Kristine Ann Peshek, Hearing Board of the Illinois Attorney Registration and Disciplinary Commission (Aug. 25, 2009), www.iardc.org/09CH0089CM.html.

35 Ibid.

36 Ibid.

37 In re Kristine Ann Peshek, Hearing Board of the Illinois Attorney Registration and Disciplinary Commission (Aug. 25, 2009), www.iardc.org/09CH0089CM.html.

38 Robert J. Ambrogi, "Lawyer Faces Discipline over Blog Post," Sep. 11, 2009, http://legalblogwatch. typepad.com/legal_blog_watch/2009/09/lawyer-faces-discipline-over-blog-posts.html; In re Kristine Ann Peshek, M.R.23794 (Ill. 2010), www.iardc.org/rd_database/disc_decisions_detail.asp.

39 Urmee Khan, "Juror Dismissed from a Trial After Using Facebook to Help Make a Decision," Nov. 24, 2008, www.telegraph.co.uk/news/newstopics/lawreports/3510926/Juror-dismissed-from-a-trial -after-using-Facebook-to-help-make-a-decision.html.

40 Ibid.

41 Paul Elias, "Courts Finally Catching Up to Texting Jurors," The Seattle Times, March 6, 2010, http:// seattletimes.nwsource.com/html/nationworld/2011274084_apustextingjurors.html.

42   Ginny LaRoe, "Barry Bonds Trial May Test Tweeting Jurors," Feb. 16, 2011, www.dailybusinessreview .com/PubArticleDBR.jsp?id=1202482052281&hbxlogin=1.

43   Erin McClam, "Stewart Uses Web to Garner Support; Marthatalks.com Tells Her Side," *Toronto Star*, Jan. 13, 2004.

44   Brian Grow, "Juror Could Face Charges for Online Research," Jan. 19, 2011, www.reuters.com /article/2011/01/19/us-internet-juror-idUSTRE70I5KD20110119.

45   Ibid.

46   *Wardlaw v. State*, 971 A.2d 331 (Md. Ct. Spec. App. 2009). A jury sitting in the Circuit Court for Baltimore City convicted Zarzine Wardlaw of rape in the second degree, sexual offense in the third degree, sexual offense in the fourth degree, three counts of assault in the second degree, two counts of sexual child abuse, and two counts of incest with his 17-year-old daughter, Michelle.

47   Ibid.

48   *Sharpless v. Sim*, 209 S.W.3d 825, 828 (Tex. Ct. App. 2006).

49   *Tapanes v. State*, 43 So. 3d 159 (Fla. Dist. Ct. App. 2010).

50   Ibid.

51   *U.S. v. Bonds*, No. 07-CR-00732 (9th Cir. 2011), Preliminary Jury Instructions, 6.

52   Brian Grow, "As Jurors Go Online, U.S. Trials Go off Track," Dec. 8, 2010, www.reuters.com /article/2010/12/08/internet-jurors-idUSN0816547120101208.

53   Ibid.

54   David Chartier, "Juror's Twitter Posts Cited in Motion for Mistrial," March 15, 2009, http:// arstechnica.com/web/news/2009/03/jurors-twitter-posts-cited-in-motion-for-mistrial.ars.

55   Ibid.

56   Ibid.

57   Grow, "As Jurors Go Online, U.S. Trials Go off Track."

58   Ibid.

59   American College of Trial Lawyers, "Jury Instructions Cautioning Against Use of the Internet and Social Networking," September 2010, www.actl.com/AM/Template.cfm?Section=Home&template= /CM/ContentDisplay.cfm&ContentID=5213.

60   Ibid.

61   LaRoe, "Barry Bonds Trial May Test Tweeting Jurors."

62   Grow, "As Jurors Go Online, U.S. Trials Go off Track."

63   Stephanie Francis Ward, "Tweeting Jurors to Face Jail Time With New California Law," Aug. 8, 2011, www.abajournal.com/news/article/tweeting_jurors_to_face_jail_time_with_new_california_law/.

12. The Right to a Fair Trial

1   "Top 10 People Caught on Facebook," www.time.com/time/specials/packages/article/0,28 804,1943680_1943678_1943557,00.html; Edward Marshall, "Burglar Leaves his Facebook Page on Victim's Computer," Sep. 16, 2009, www.journal-news.net/page/content.detail/id/525232.html.

2   Carl Campanile, "Dem Pol's Son was 'Hacker,'" Sep. 19, 2008, www.nypost.com/p/news/politics /item_IJuyiNfQAkPKvZxRXvSEyJ;jsessionid=E00BF6768C23A24BD95F4E906DDD74B7; Bill Poovey, "David Kernell, Palin E-mail Hacker, Sentenced to Year in Custody," Nov. 12, 2010, www.huffingtonpost .com/2010/11/12/david-kernell-palin-email_n_782820.html.

3   He also posted screenshots from the email account. He was sentenced to a year in jail, but the judge changed it to a term in a halfway house. *U.S. v. Kernell*, 742 F. Supp. 2d 904 (E.D. Tenn. 2010); Poovey, "David Kernell, Palin Email Hacker, Sentenced to Year in Custody."

4   Robert Quigley, "How the Social Web Snared Yale Murder Suspect Raymond Clark," Sep. 17, 2009, www.mediaite.com/online/how-the-web-of-social-media-snared-yale-murder-suspect-raymond-clark/.

5   Los Angeles Police Department, "How Gangs Are Identified," www.lapdonline.org/get_informed /content_basic_view/23468.

6   *State v. Trusty*, 776 N.W.2d 287 (Wis. Ct. App. 2009).

7   IACP, Center for Social Media Survey Results, September 2010, at 3, www.iacpsocialmedia.org /Portals/1/documents/Survey%20Results%20Document.pdf.

8   Ibid. at 9.

9   "Petrick Googled 'Neck,' 'Snap,' Among Other Words, Prosecutor Says," Nov. 13, 2005, updated Dec. 10, 2006, www.wral.com/news/local/story/121729/.

10   *U.S. v. Taylor*, 956 F.2d 572, 590 (6th Cir. 1992) (Martin, J., dissenting).

11   Kary L. Moss, "Substance Abuse During Pregnancy," 13 *Harvard Women's Law Journal* 278, 294 (1990).

12   Erika L. Johnson, "'A Menace to Society': The Use of Criminal Profiles and Its Effect on Black Males," 38 *Howard Law Journal* 629, 641 (1995).

13   Jake Griffin, "Teen Drinkers Spotted on Web," *Daily Herald* (Arlington Heights, Ill.), Jan. 12, 2008, at 1.

14   Trench Reynolds, "Conviction in MySpace Weapons Posting," April 5, 2006, http://trenchreynolds .me/?s=father+arrested+for+son+posing+with+gun.

15   Caroline Black, "'Baby with Bong' Facebook Photo Arrest; Fla. Mom Rachel Stieringer Faces Drug Charges," Aug. 17, 2010, www.cbsnews.com/8301-504083_162-20013878-504083.html.

16   "Baby Bong Photo on Facebook Lands Mother in Jail," Aug. 18, 2010, www.pakpoint.net/baby -bong-photo-on-facebook-lands-mother-in-jail/16099/. She has since entered a four-month pretrial intervention program; "Mom of Bong Baby in Pretrial Intervention," Feb. 16, 2011, www.news4jax .com/news/27037785/detail.html.

17   "Pair Accused of Catching, Eating Iguana," *Sun Sentinel*, Feb. 9, 2009, at 12A.

18   Ibid.

19   "Ulrich Mühe" (obituary), July 27, 2007, www.telegraph.co.uk/news/obituaries/1558608/Ulrich -Muhe.html.

20   Ibid.

21   *State v. Greer*, 2009 WL 2574160 (Ohio Ct. App. 2009).

22   Daniel Findlay, "Tag! Now You're Really 'It' What Photographs on Social Networking Sites Mean for the Fourth Amendment," 10 *North Carolina Journal of Law and Technology* 171 (2009); Associated Press, "Facebook Evidence Sends Unrepentant Partier to Prison," July 21, 2008, www.foxnews.com /printer_friendly_story/0,3566,386241,00.html.

23   Associated Press, "Facebook Evidence Sends Unrepentant Partier to Prison."

24   Ibid.

25   *Williamson v. State*, 2011 Ark. App. 73 (Ark. Ct. App. 2011).

26   *People v. Beckley*, 185 Cal. App. 4th 509 (Cal. Ct. App. 2010).

27   *In re K.W.*, 666 S.E.2d 490, 494 (N.C. App. 2008).

28   *U.S. v. Jackson*, 208 F.3d 633 (7th Cir. 2000).

29   *State v. Bell*, 882 N.E.2d 502, 508 (Ohio Ct. Com. Pl. 2008).

30   "DA: MySpace Used in Jury Tampering," Aug. 26, 2009, www.kcra.com/r/20571181/detail.html.

31   "DA: Woodland Woman Tracked Juror on MySpace, Pleaded for Boyfriend's Acquittal," Aug. 26, 2009, www.news10.net/news/local/story.aspx?storyid=65888.

32 "DA: MySpace Used in Jury Tampering."

33 "DA: Woodland Woman Tracked Juror on MySpace, Pleaded for Boyfriend's Acquittal."

34 Ibid.

35 "Local Low-Life Convicted for Domestic Violence, Victim Accused of Jury Tampering," Aug. 27, 2009, http://woodlandrecord.com/local-lowlife-convicted-for-domestic-violence-victim-accused -of-jury-tamp-p831-1.htm.

36 Ibid.

37 *Griffin v. State*, 995 A.2d 791 (Md. Ct. Spec. App. 2010).

38 Ibid.

39 Ibid., *citing Lorraine v. Markel American Insurance Company*, 241 F.R.D. 534, 569 (D. Md. 2007).

40 *People v. Heeter*, 2010 WL 2992070 (Cal. Ct. App. 2010), unpublished/noncitable (Aug. 2, 2010), review denied (Cal. Nov. 17, 2010).

41 Ibid.

42 Bethany Krajelis, "Social Media Consulting Becomes Part of Trial Strategy," July 28, 2011, www .akronlegalnews.com/editorial/1050; Sylvia Hsieh, "On Murder and Social Media: Casey Anthony's Jury Consultant Speaks," July 5, 2011, http://lawyersusaonline.com/blog/2011/07/05/on-murder -and-social-media-casey-anthony%E2%80%99s-jury-consultant-speaks/; Walter Pacheco, "Casey Anthony: How Social Media Tweaked Defense Strategy," July 13, 2011, http://articles.orlandosentinel .com/2011-07-13/business/os-casey-anthony-social-media-strateg20110713_1_media-sites-casey -anthony-cindy-anthony; Julie Kay, "Did Social Media Make Casey Anthony's Case?," July 12, 2011, www.law.com/jsp/lawtechnologynews/PubArticleLTN.jsp?id=1202500225278&Did_Social_Media _Make_Casey_Anthonys_Case.

43 Hsieh, "On Murder and Social Media."

44 WESHTV, "Expert: Chloroform Page Accessed 84 Times," June 8, 2011, www.youtube.com /watch?v=s5GvkDRWRLM.

45 WESHTV, "Day 24: Defense Witness Backfires, Baez Sanctioned," June 21, 2011, www.youtube .com/watch?v=vpapMdDvn7E.

46 Lizette Alvarez, "Software Designer Reports Error in Anthony Trial," *The New York Times*, July 19, 2011, at A14, www.nytimes.com/2011/07/19/us/19casey.html?_r=1&scp=1&sq=casey%20 anthony%20Bradley%20computer%20expert&st=cse.

47 Judge Gena Slaughter and John G. Browning, "Social Networking Do's and Don'ts for Lawyers and Judges," 73 *Texas Bar Journal* 192, 194 (2010).

48 Ibid. at 194.

13. The Right to Due Process

1 Electronic Privacy Information Center's Complaint to the Federal Trade Commission at 22, *In the Matter of Facebook, Inc. and the Facial Identification of Users* (June 10, 2011); "Facebook Asks More than 350 Million Users Around the World to Personalize Their Privacy," Dec. 9, 2009, www .facebook.com/press/releases.php?p=133917.

2 Complaint, Request for Investigation, Injunction, and Other Relief, before the Federal Trade Commission, *In the Matter of Facebook, Inc.* (Dec. 17, 2009) http://epic.org/privacy/inrefacebook /EPIC-FacebookComplaint.pdf, *citing* Farnaz Fassihi, "Iranian Crackdown Goes Global," Dec. 3, 2009, http://online.wsj.com/article/SB125978649644673331.html.

3 Ibid. *citing* Fassihi, "Iranian Crackdown Goes Global."

4 Miguel Helft and Brad Stone, "With Buzz, Google Plunges into Social Networking," *The New York Times*, Feb. 10, 2010, at B3, www.nytimes.com/2010/02/10/technology/internet/10social.html.

5   Complaint by the Federal Trade Commission at ¶ 8, *In the Matter of Google Inc.* (2010).

6   Ibid. at ¶ 9.

7   Ibid. at ¶¶ 7, 9.

8   Ibid. at ¶ 9.

9   Ibid. at ¶ 10.

10   Byron Acohido, "FTC Slaps Google with Audits over Buzz," March 31, 2011, www.usatoday.com /tech/news/2011-03-30-google-ftc-settlement.htm?csp=34tech; Robin Wauters, "Google Buzz Privacy Issues Have Real Life Implications," Feb. 12, 2010, http://techcrunch.com/2010/02/12/google-buzz -privacy/.

11   Complaint, *Lane v. Facebook, Inc.*, 2008 WL 3886402 (N.D. Cal. Aug. 12, 2008).

12   Ibid.

13   Ellen Nakashima, "After Spoiling Holiday Surprises, Facebook Changes Ad Feature," *The Washington Post*, Nov. 30, 2007, at A01, published online as "Feeling Betrayed, Facebook Users Force Site to Honor Their Privacy," www.washingtonpost.com/wp-dyn/content/article/2007/11/29 /AR2007112902503.html.

14   Ibid.

15   Jon Brodkin, "Facebook Halts Beacon, Gives $9.5M to Settle Lawsuit," Dec. 2009, www.pcworld .com/article/184029/facebook_halts_beacon_gives_95m_to_settle_lawsuit.html.

16   Settlement Agreement at 8, *Lane v. Facebook, Inc.*, No. 08-CV-3845 RS (N.D. Cal. Sep. 17, 2009); Plaintiff's Notice of Motion and Motion for Preliminary Approval of Class Action Settlement Agreement at 8, *Lane v. Facebook, Inc.*, 2009 WL 3169921 (N.D Cal. Sep. 18, 2009).

17   Guilbert Gates, "Facebook Privacy: A Bewildering Tangle of Options," May 12, 2010, www.nytimes.com /interactive/2010/05/12/business/facebook-privacy.html.

18   "7,500 Online Shoppers Unknowingly Sold Their Souls," April 15, 2010, www.foxnews.com /scitech/2010/04/15/online-shoppers-unknowingly-sold-souls/?test=latestnews.

19   *Connally v. General Construction Co.*, 269 U.S. 385, 392 (U.S. 1926).

20   Electronic Privacy Information Center's Complaint to the Federal Trade Commission at 21, *In the Matter of Facebook, Inc. and the Facial Identification of Users* (June 10, 2011).

21   Ibid. at 21.

22   Ibid. at 21, *citing* Justin Smith, "Scared Students Protest Facebook's Social Dashboard, Grappling with Rules of Attention Economy," Sep. 6, 2006, www.insidefacebook.com/2006/09/06/scared -stuents-protest-facebooks-social-dashboard-grappling-with-rules-of-attention-economy/.

23   Ibid. *citing* Mark Zuckerberg, "An Open Letter from Mark Zuckerberg," Sep. 8, 2006, http://blog .facebook.com/blog.php?post=2208562130.

24   *Specht v. Netscape Communications Corp.*, 306 F.3d 17, 32 (2d Cir. 2002).

25   Required Warnings for Cigarette Packages and Advertisements, 76 Fed. Reg. 36628 (June 22, 2011) (to be codified at 21 C.F.R. pt. 1141).

26   "Consumer Reports Poll: Americans Extremely Concerned About Internet Privacy," Sep. 25, 2008, http://consumersunion.org/pub/core_telecom_and_utilities/006189.html.

27   "About us," http://chitika.com/blog/about-us/.

28   Complaint at ¶ 9, *In the Matter of Chitika, Inc.*, 2011 WL 914035 (F.T.C. March 14, 2011), www.ftc .gov/os/caselist/1023087/110314chitikasmpt.pdf.

29   Decision and Order, *In the Matter of Chitika, Inc.*, 2011 WL 2487158 (F.T.C. June 7, 2011), www .ftc.gov/os/caselist/1023087/110314chitikasmpt.pdf.www.ftc.gov/os/caselist/1023087/110617chitika do.pdf.

30 "Online Privacy, Social Networking and Crime Victimization: Hearing Before the Subcomm. on Crime, Terrorism and Homeland Sec. of the H. Comm. on the Judiciary," statement of Marc Rotenberg, Executive Director, Electronic Privacy Information Center, 111th Cong. 5, 2010.

31 *Mathews v. Eldridge*, 424 U.S. 319 (1976).

32 First Amended Complaint at ¶ 31, *Robins v. Spokeo*, No. 10-CV-05306 (C.D. Cal. Feb. 16, 2011).

33 Ibid.

34 15 U.S.C. §1681j(a)(1)(C).

35 Linda A. Szymanski, "Sealing/Expungement/Destruction of Juvenile Court Records: Records That Can Be Sealed or Inspected," 11 *National Center for Juvenile Justice Snapshot* (2006), www.ncjj.org /PDF/Snapshots/2006/vol11_no11_recordsunsealedinspected.pdf.

36 George Blum, Romualdo P. Eclavea, Alan J. Jacobs, John Kimpflen, Jack K. Levin, Caralyn M. Ross, Jeffrey J. Shampo, Eric C. Surette, Amy G. Gore, Glenda K. Harnad, John R. Kennel, and Mary Babb Morris, "Circumstances Under Which Expungement Can Be Ordered," 21A *American Jurisprudence, Second Edition* §1222 (2011).

37 Viktor Mayer-Schönberger, *Delete: The Virtue of Forgetting in the Digital Age* (Princeton, N.J.: Princeton University Press, 2009), 171.

38 Mass. H. 02705, 187th Gen. Ct. (2011).

39 Mass. H. 02705, 187th Gen. Ct. §7 (2011).

40 European Union Directive 95/46/EC, Section I, Article 6.

41 Joel Stein, "Your Data, Yourself," *Time Magazine*, March 21, 2011, at 40, published online on March 10, 2011, as "Data Mining: How Companies Now Know Everything About You," www.time.com /time/magazine/article/0,9171,2058205-1,00.html.

42 Richard Power, "Face Recognition and Social Media Meet in the Shadows," CSO Security and Risk, Aug. 1, 2011, www.csoonline.com/article/686959/face-recognition-and-social-media-meet-in-the -shadows.

43 The Nuremberg Code of Ethics in Medical Research.

44 Maija Palmer, "Hamburg Rules Against Facebook Facial Recognition," *Financial Times*, Aug. 3, 2011, at 12.

45 Beth Wellington, "What Facebook Fails to Recognise," June 14, 2011, www.guardian.co.uk /commentisfree/cifamerica/2011/jun/14/facebook-facial-recognition-software .

46 Nick Bilton, "Facebook Changes Privacy Settings to Enable Facial Recognition," The New York Times Bits blog, June 7, 2011, http://bits.blogs.nytimes.com/2011/06/07/facebook-changes-privacy -settings-to-enable-facial-recognition/?hp.

47 Ibid.

48 Electronic Privacy Information Center's Complaint to the Federal Trade Commission at 12, *In the Matter of Facebook, Inc. and the Facial Identification of Users* (June 10, 2011).

49 Justin Mitchell, "Making Photo Tagging Easier," June 9, 2011, www.facebook.com/blog .php?post=467145887130.

50 Electronic Privacy Information Center's Complaint to the Federal Trade Commission at 17, *In the Matter of Facebook, Inc. and the Facial Identification of Users* (June 10, 2011).

51 Ibid. at 1.

52 The Hamburg Data Protection Authority is responsible for supervising compliance with data protection laws in one of Germany's 16 states. *See*, "Data Protection Authorities in Germany," www .ldi.nrw.de/LDI_EnglishCorner/mainmenu_AboutLDI/index.php.

53 "Facebook Violates German Law, Hamburg Data Protection Official Says," Aug. 2, 2011, www .dw-world.de/dw/article/0,,15290120,00.html.

54   Maija Palmer, "Hamburg Rejects Facebook Facial Recognition," Aug. 2, 2011, www.ft.com/cms/s
      /0/14007238-bd29-11e0-9d5d-00144feabdc0.html#ixzz1TzhMD4ha.

55   Stephanie Bodoni, "Facebook to Be Probed in EU for Facial Recognition in Photos," June 8, 2011,
      www.businessweek.com/news/2011-06-08/facebook-to-be-probed-in-eu-for-facial-recognition-in
      -photos.html.

56   Jenna Greene, "Google Settles FTC Charges, Agrees to Boost Privacy Protections," March 30, 2011,
      http://legaltimes.typepad.com/blt/2011/03/google-settles-ftc-charges-agrees-to-bost-privacy-protections
      -.html.

57   Agreement Containing Consent Order, *In the Matter of Google Inc.*, 2011 WL 1321658 (F.T.C.
      March 30, 2011).

58   Byron Acohido, "FTC Slaps Google with Audits over Buzz," March 31, 2011, http://abcnews.go.com
      /Technology/ftc-slaps-google-audits-buzz/story?id=13262319.

59   Settlement Agreement at 8, *Lane v. Facebook, Inc.*, No. 08-CV-3845 RS (N.D. Cal. Sep. 17, 2009).

60   www.beaconclasssettlement.com/FAQs.htm#FAQ3.

61   Settlement Agreement at 11, *Lane v. Facebook, Inc.*, No. 08-CV-3845 RS (N.D. Cal. Sep. 17, 2009).

62   Reply Brief for Appellant Ginger McCall, *Lane v. Facebook, Inc.*, No. 10-16398 (9th Cir. Jan. 18,
      2011).

63   Wendy Davis, "Facebook's Beacon Settlement Appealed by Privacy Advocate," June 24, 2010, www
      .mediapost.com/publications/?fa=Articles.showArticle&art_aid=130897.

64   Electronic Privacy Information Center's Complaint to the Federal Trade Commission, *In the Matter
      of Facebook, Inc. and the Facial Identification of Users* (June 10, 2011).

65   Naomi Klein, "China's All-Seeing Eye," May 14, 2008, www.naomiklein.org/articles/2008/05/chinas
      -all-seeing-eye.

66   Keith Bradsher, "Theft Reveals Lapses in Chinese Museum's Security," May 12, 2011, www.nytimes
      .com/2011/05/13/world/asia/13beijing.html?_r=1.

67   "Face Recognition: Anonymous No More," *The Economist*, July 30, 2011, at 51, www.economist
      .com/node/21524829.

14. Slouching Towards a Constitution

1   Cory Doctorow, Talk at TEDx Observer, 2011, http://tedxtalks.ted.com/video/TEDxObserver-Cory
      -Doctorow.

2   Ibid.

3   M.G. Siegler, "Google+ Project: It's Social, It's Bold, It's Fun, and It Looks Good—Now for the Hard
      Part," June 28, 2011, http://techcrunch.com/2011/06/28/google-plus/.

4   Claire Cain Miller, "Google(Plus) Looks a Lot like Facebook," June 29, 2011, www.timesunion.com
      /business/article/Google-Plus-looks-a-lot-like-Facebook-1444698.php#ixzz1RLyU9z5F.

5   Claire Cain Miller, "Another Try by Google to Take on Facebook," *The New York Times*, June 28,
      2011, at B1, www.nytimes.com/2011/06/29/technology/29google.html?_r=2&pagewanted=1.

6   Declan McCullagh, "Google+ Steers Clear of Privacy Missteps," June 29, 2011, http://news.cnet
      .com/8301-31921_3-20075281-281/google-steers-clear-of-privacy-missteps/#ixzz1RGNxdx5r.

7   David Lazarus, "Forget Privacy; He Sells Your Data," *Los Angeles Times*, June 8, 2010, at 1, http://
      articles.latimes.com/2010/jun/08/business/la-fiw-lazarus-20100608.

8   "2009 Study: Consumer Attitudes About Behavioral Targeting," 2009, www.truste.com/pdf/TRUSTe
      _TNS_2009_BT_Study_Summary.pdf.

9   Joseph Turow, Jennifer King, Chris Jay Hoofnagle, Amy Bleakley, and Michael Hennessy,
      "Contrary to What Marketers Say, Americans Reject Tailored Advertising and Three Activities

That Enable It," Federal Trade Commission, September 2009, at 24, www.ftc.gov/os/comments/privacyroundtable/544506-00113.pdf.

10 Ibid.

11 Martha Irvine, "Who Develops Social Networking Conscience? Youth, Not Elders," May 31, 2010, http://wraltechwire.com/business/tech_wire/news/blogpost/7697690/.

12 Viktor Mayer-Schönberger, *Delete: The Virtue of Forgetting in the Digital Age* (Princeton, N.J.: Princeton University Press, 2009), 137.

13 Directive 95/46/EC of the European Parliament and of the Council, Oct. 24, 1995, 11.

14 "Data Protection in the European Union," last updated Feb. 24, 2011, at 9, http://ec.europa.eu/justice/policies/privacy/guide/index_en.htm.

15 Jeff Jarvis, "A Bill of Rights in Cyberspace," March 27, 2010, www.buzzmachine.com/2010/03/27/a-bill-of-rights-in-cyberspace/.

16 "It's Time for a Social Network Users' Bill of Rights," Computers, Freedom, and Privacy blog, June 6, 2010, http://cfp.org/wordpress/?p=341.

17 The Computers, Freedom, and Privacy Conference proposed this Bill of Rights:

Honesty: Honor your privacy policy and terms of service.

Clarity: Make sure that policies, terms of service, and settings are easy to find and understand.

Freedom of speech: Do not delete or modify my data without a clear policy and justification.

Empowerment: Support assistive technologies and universal accessibility.

Self-protection: Support privacy-enhancing technologies.

Data minimization: Minimize the information I am required to provide and share with others.

Control: Let me control my data, and don't facilitate sharing it unless I agree first.

Predictability: Obtain my prior consent before significantly changing who can see my data.

Data portability: Make it easy for me to obtain a copy of my data.

Protection: Treat my data as securely as your own confidential data unless I choose to share it, and notify me if it is compromised.

Right to know: Show me how you are using my data and allow me to see who and what has access to it.

Right to self-define: Let me create more than one identity and use pseudonyms. Do not link them without my permission.

Right to appeal: Allow me to appeal punitive actions.

Right to withdraw: Allow me to delete my account and remove my data.

18 Mayer-Schönberger, *Delete: The Virtue of Forgetting in the Digital Age*, 111.

INDEX

Page numbers in *italics* refer to charts.

ABOUT THE AUTHOR

Lori Andrews is a law professor and the director of the Institute for Science, Law and Technology at the Illinois Institute of Technology. A frequent guest on *Oprah*, *60 Minutes*, *Today*, and *Nightline*, she was named a "Newsmaker of the Year" by the *American Bar Association Journal* and one of the 100 Most Influential Lawyers in America by the *National Law Journal*. She has served as a regular advisor to the U.S. government on new technologies, including chairing the federal committee concerning ethical and legal issues involved with the Human Genome Project. Learn more at www.loriandrews.com. And vote on the Social Network Constitution at www.socialnetworkconstitution.com.